ENVIRONMENTAL SCIENCE, ENGINEERING AND TECHNOLOGY

INVASIVE SPECIES

THREATS, ECOLOGICAL IMPACT AND CONTROL METHODS

Environmental Science, Engineering and Technology

Additional books in this series can be found on Nova's website under the Series tab.

Additional e-books in this series can be found on Nova's website under the e-book tab.

Wildlife Protection, Destruction and Extinction

Additional books in this series can be found on Nova's website under the Series tab.

Additional e-books in this series can be found on Nova's website under the e-book tab.

ENVIRONMENTAL SCIENCE, ENGINEERING AND TECHNOLOGY

INVASIVE SPECIES

THREATS, ECOLOGICAL IMPACT AND CONTROL METHODS

JOAQUÍN J. BLANCO
AND
ADRIAN T. FERNANDES
EDITORS

Nova Science Publishers, Inc.
New York

For permission to use material from this book please contact us:
Telephone 631-231-7269; Fax 631-231-8175
Web Site: http://www.novapublishers.com

Library of Congress Cataloging-in-Publication Data

ISBN: 978-1-61942-761-7

Published by Nova Science Publishers, Inc. ✝ New York

CONTENTS

PREFACE

Many species of plants, microbes, and animals have been introduced around the world. Species are considered "invasive" or "alien" when they are not native to an area, but become established and cause, or are likely to cause, economic or environmental harm or harm to human health. In this book, the authors present current research in the study of the threats, ecological impacts and control methods relating to invasive species. Topics include thermochemical processing of wood from invasive arboreal species for sustainable bioenergy generation; invasive rodents in the U.S.; invasive Buprestid beetle species monitoring; the invasive beaver castor canadensis of the Tierra del Fuego Archipelago; leaf-cutting ants; invasive species and accurate ecological assessment in freshwater pollution; and the aquarium trade as a pathway for the introduction of invasive species in Mexico.

Chapter 1 – Thermochemical conversion of wood sawdust from two invasive arboreal species (*Prosopis ruscifolia* and *Parkinsonia aculeata)* into valuable products is explored to provide novel possible means of spread control attempting to their use as alternative, easily renewable bio-resources for bioenergy generation and activated carbons production. Rapid pyrolysis of both wood species is examined to assess the feasibility of their conversion into energy products. Yields of the three kinds of pyrolysis products, comprising bio-oil, non-condensable gas, and bio-char, are determined from experiments performed in a bench-scale installation at 500 °C. The wood species noticeably affects yields and distributions of the pyrolysis products. Main physicochemical characteristics and higher heating value (HHV) of the products generated from *Prosopis ruscifolia*, leading to higher bio-oil yield (~42%) than *Parkinsonia aculeata*, are determined. All the products show potentialities for energy applications. The bio-oil, further upgraded, could be employed as a liquid fuel with characteristics similar to those of fuel-oil, whereas the low heating value-gas evolved with pyrolysis course, mostly composed by CO_2, CO, CH_4 and H_2, might contribute to energy sustainability of the process. In turn, the bio-char (HHV≈ 27 MJ/kg) has potential as environmentally friendly solid bio-fuel. Kinetic measurements for the pyrolysis of the two wood species are also carried out by non-isothermal thermogravimetric analysis from room temperature up to 500 °C. An overall first-order decomposition model enables to properly represent kinetic data for both species over the whole temperature range. Differences in pyrolytic reactivity are reflected in the estimated kinetic parameters. On the other hand, use of wood sawdust from the two species as precursor of activated carbons targeted at wastewater treatment is investigated. For this purpose, the phosphoric acid activation process is applied at pre-established moderate conditions (acid/precursor weight ratio of 2, 450 ºC, 0.5 h). The activation atmosphere involved in the thermal treatment stage of the process is *in situ* modified to examine its influence on the development of porous structures, surface chemistry, and adsorptive behaviour of the resulting activated carbons. Either a self-generated atmosphere, flowing air or nitrogen is used. The activation atmosphere and the wood species have a marked influence on the characteristics of the derived carbons. Those developed from *Prosopis ruscifolia* wood in the self-generated atmosphere show higher BET surface area and total pore volume (2281 m^2/g, 1.7 cm^3/g) than the ones obtained under flowing air (1638 m^2/g and 1.3 cm^3/g), although the latter possess a higher total content of surface acidic/polar oxygen groups (2.2 meq/g) than the former (1.5 meq/g). Textural characteristics of *P. ruscifolia*-based activated carbons are superior to those for the carbons developed from *P. aculeata* wood in the self-generated atmosphere (968 m^2/g, 0.70 cm^3/g) and flowing N_2 (1103 m^2/g, 0.75 cm^3/g). These samples have total contents of surface acidic/polar oxygen functionalities of 1.9 meq/g and 1.7 meq/g, respectively. The ability of the activated carbons in removing representative toxic metal ions, oxo-anions, and organic compounds is examined from batch assays using dilute aqueous solutions as models of wastewater. Adsorption isotherms of Cu[II] ion onto *P. ruscifolia*-based activated carbons are determined and properly described by the Langmuir model. Maximum sorption capacity (X_m) for the air-derived carbons (X_m= 0.44 mmol/g) almost duplicates the value for those obtained in the self-generated atmosphere (X_m= 0.24 mmol/g), pointing to a predominant effect of the surface acidic functionalities on metal sequestering behaviour. Adsorption isotherms of salicylic acid onto the air-derived carbons are also ascertained, and satisfactorily fitted to the Langmuir model. On the other hand, effectiveness of the *P. aculeata*-based activated carbons in removing individual nitrate ion, phenol, and methylene blue at pre-established equilibrium conditions is comparatively investigated. High removal levels are attained for the organics (~ 90% and 100% for phenol and methylene blue, respectively), but nitrate removal is low ($\leq$ 40%). Removal effectiveness of these carbons is inversely related to water solubility of the solutes and is slightly affected by the activation atmosphere. Accordingly, effective activated carbons of potential applicability in tertiary wastewater treatment may be developed from the two invasive species, as suitable, alternative precursors, by phosphoric acid activation at moderate conditions.

Chapter 2 – Many invasive rodents have become established in the United States and its territories. The species include several species of *Rattus*, house mice (*Mus musculus*), Gambian giant pouched rats (*Cricetomys gambianus*), ground squirrels (*Spermophilus parryii*), nutria (*Myocastor coypus*) and marmots (*Marmota caligata*). While most were introduced accidently, some were introduced for food or fur. Additionally, some native species of rodents have been placed on islands, at least on a temporary basis, to study rodent species interactions. These rodents have caused serious impacts to native flora and fauna, agriculture, and other resources. They have caused the extinction or many species of birds on insular ecosystems. Although many methods are used to control or eradicate introduced rodents, rodenticides and traps are the main tools. Since the early 1990s, agencies have been eradicating rodents from various islands, primarily for conservation purposes. Of about 27 eradication attempts, 22 (81%) appear to have succeeded with only about 5 failures. For several islands, however, it is too early to determine if the attempted eradication has been successful or not. In the case of failed eradications, rapid re-invasion by rodents from nearby islands may be the reason. Numerous additional eradications are planned. We review the introduced rodent species, their impacts, and eradications, both successful and unsuccessful, that have occurred in the United States. Most eradications involved the use of the anticoagulant rodenticides diphacinone and brodifacoum. Rodenticides have been applied by hand-broadcast, bait station deployment, and aerial broadcast. We briefly review the strategies and methods used in eradication projects and the efforts to mitigate potential non-target and environmental impacts. Finally, we consider some of the remaining challenges in invasive rodent management and eradication in the United States, including the use of toxicants, land access, public attitudes, resource availability and monitoring difficulties.

Chapter 3 – Buprestid beetles (Coleoptera: Buprestidae), which were previously considered to be minor forest pests, have received substantial attention recently due to the widespread ash tree mortality associated with the North America introduction of the emerald ash borer, *Agrilus planipennis* from Asia. At the same time, geographic expansions of forest buprestid species associated with other hosts have been noticed, causing concern about potentially similar destructive outcomes. Attempts have been made to understand the host-finding and mating behaviors of these forest buprestids in order to develop detection and management tools, with a focus primarily on *A. planipennis*. A complex sequence of chemical and visual signals appears to mediate such behaviors, which has been difficult to mimic in an effective trapping approach. Both sexes appear to be attracted to the tree canopy as a result of host-produced volatile compounds. Once in the canopy, males visually seek and fly directly onto stationary females from a distance of up to 1 meter, an event which immediately precedes attempts to copulate. The subsequent copulation attempts by the males are mediated by sex-specific cuticular hydrocarbon signals. Similar visually-mediated approaches toward dead models have also been observed in other tree-dwelling buprestids such as *Agrilus cyanescens*, *A. subcinctus*, *A. biguttattus*, *A. sulcicollis*, and *A. angustulus*. While the extent of similarities of both visual and chemical signals shared by these species and *A. planipennis* has not been fully explored, it is becoming clear that the visual signal that induces attraction to a female is broadly tuned, exhibiting substantial cross-species attraction. Despite subtle differences in coloration patterns, all of these species generate an intense spectral signal using the refractive properties of repetitive layering of the elytral cuticle. Further investigation into both the means of generation and response to the visual signal, as well as the chemical ecology of these species, may provide advances in trapping technologies for the detection of all tree-feeding Buprestids.

Chapter 4 – The Tierra del Fuego Archipelago (TDFA) contains numerous invasive species, of which *Castor canadensis* is the most abundant and important invasion. Beaver are responsible for the most drastic landscape alteration in Tierra del Fuego since the last glacial age, affecting not only the hydrology and composition of the southern beech forest, but more importantly allowing other exotic species to invade the ecosystem. From 25 pairs intentionally released in 1946, beavers have increased their numbers to a current population size of aprox.100,000 individuals. Here, we present the genetic variability and population structure of 222 mitochondrial DNA beaver samples. We detected 7 D-loop haplotypes, 3 of them resulted the most abundant and distributed all along the archipelago. To analyze the data, TDFA population was subdivided into five subpopulations and a slight structure was found (ϕst =0.169). With these data we also identify management units (MUs) of the invasive beaver in the TDFA. Although, sea water and mountains could be considered barriers for beaver's dispersion, we found no barriers to gene flow within the Isla Grande, which may be due to the complex stream watershed network that allows beavers to easily disperse. These results indicate that the Isla Grande is a single management unit (MU) and every small island in the archipelago is a separate MU. It is important to consider that successful eradication programs of mammals always were conducted in small islands while invasive populations on larger islands or those that display no distinct structure are more problematic, so the eradication of *Castor canadensis* in the TDFA would be logistically very difficult or perhaps impossible. Therefore, we propose to (1) control the expansion from IG population to others areas through trapping beavers and (2) eradicate beaver populations of the adjacent small islands (small area comparative to Isla Grande) and especially from the continent. If the species is not early removed in the mainland, beavers will begin to invade the Patagonian forests very quickly and its future eradication in these continental areas will be almost impossible. To carry out the control of the species in Tierra del Fuego, is necessary to promote beaver extraction through the commercial trapping in a way to control this species in the short term. To achieve this important point, the classic control scheme must be change it stimulating new business development by incorporating new materials and products to market and also coordinate the participation of public and private actions to exploit species under control. In this work, we state three options for economic use of the beaver: 1) the government could pay a price for killing beavers, 2) to sell/export the beaver fur and 3) to sell beaver′s meat for human consumption. With these three options of beaver exploitation, we believe that the Isla Grande control may be feasible and logistically possible.

Chapter 5 – The fungus-growing ants belong to the genus Atta and Acromyrmex and individuals are causing serious damage to various types of plantations due to the habit of cutting leaves and shoots of economic importance for use in cultivating the garden symbiotic fungus colonies, causing serious damage. The ants attack is harmful at any stage of the cycle, but the damage is greater in the training phase of the plant, when temporarily paralyzes growth. In an attempt to reduce the damage caused by these insects farmers make the control of fire ants through the use of mechanical methods, biological and chemical. The mechanical control is to dig the nest to remove the queen (individual responsible for placing the eggs that give rise to other individuals in the colony), which are larger in size than workers, causing a disruption of the nest, since the queen controls all activities and individuals in the colony. This type of control has been recommended only for newly formed nests, otherwise, the queen will be housed in great depth, making its removal. It should be noted that in practice such control is not a good alternative in commercial plantations due to the large extent of the property. Chemical control can be accomplished by applying powder, liquid insecticide, thermonebulization and granulated bait. The powder is a type of insecticides formulated in solid vehicles, such as talc, and the application is through hand pumps. The death of ants occurs by direct contact with the chemical. However, this type of control has limitations such as the impossibility of penetration of the product in the deepest chambers of nesting adults, and has now been replaced by liquid thermonebulized, through which the product is injected into the nest with the help of hose, and toxic bait, which is considered more efficient due to easy handling, fast action and lower costs. The bait is distributed on the tracks near the colony of ants to be transported into the nest by workers. This method can cause much damage to non-target ants and even other bodies. Finally, it has been used the biological control, which is the introduction of natural enemies in plantations, such as birds, some species of spiders, beetles, mites and even other species of ants, and much research is currently focused on creating natural insecticides that have repellent action or lethal to the ants, such as castor oil, or fungicidal against the garden of fungus that feeds the colony, as is the case of sesame. Studies on the morphology of the reproductive organs of these ants could also be an important tool in the search for new methods of control. Some studies have been conducted on the morphology of the spermathecae of ants, the body that produces all the elements capable of maintaining viable sperm inside, where it was observed that products circulating in the hemolymph, due to the thin wall thickness of the the spermatheca, also could enter the environment within this structure, where are the sperm and thus affect their viability.

Chapter 6 – Macroinvertebrates are routinely used in the calculation of indices to assess the quality of rivers throughout the world. Macroinvertebrate assemblages have proved themselves to be reliable, cost effective indicators of freshwater pollution, as the relative sensitivities of their members respond to changes in water quality in predictable ways. Such biological monitoring can have advantages over chemical monitoring, as it tends to be cheaper and more reliable over the long term, especially in rivers where water quality can fluctuate dramatically on a day to day basis. However, an increasing threat to accurate biological monitoring comes from invasive species. The rate of freshwater invasions is increasing, with increasing global transport, climate change and environmental degradation all facilitating the arrival of pollution tolerant and often aggressive invaders in many previously undisturbed river systems. Such invasive species can have a major impact on benthic macroinvertebrate assemblage structure and diversity. Macroinvertebrate invaders often radically alter native macroinvertebrate assemblages through biotic interactions, such as competition and predation. In many cases more pollution tolerant invaders can replace more sensitive natives and very predatory invaders can decimate native assemblages as a whole. This has serious implications for accurate biological monitoring and the consequent assessment of the ecological status of invaded rivers. The whole structure of existing water quality indices may need revising to take into account the presence of invasive species within monitored assemblages, such that the accuracy of any derived assessments of water quality can be judged in the context of whether the assemblages generating them contain mainly natives or invaders. In addition, methods of 'biocontamination' assessment, designed to be used alongside current routine water quality indices may represent a simple, cost-effective way to integrate the biological pressure invasives represent, into a country's or region's routine river quality monitoring programme. Such methods should be adopted alongside routine biological monitoring of pollution to provide a more comprehensive estimate of ecological quality status. It is clear that he problems of invasive species and accurate monitoring of freshwater pollution can no longer be ignored.

Chapter 7 – The total ornamental fish industry worldwide (including dry goods) is valued at approximately US $15 billion dollars and it has been estimated that approximately one billion ornamental fish are commercialized every year with a value in the order of US $6 billion dollars. Freshwater species constitute the bulk of the trade; 90 percent of these are obtained from aquaculture and only 10 percent are wild captured. Around 800 to 1000 species and varieties are traded worldwide. Ornamental fish trade has been recognized as an important pathway for the introduction of non native species in several countries and present trends indicate that this pathway may turn into the main source of exotic invasive species in North America. Invasive species are characterized by posing different threats to the environment, the economy and human health. Among the main impacts provoked in the aquatic environment by invasive species are: competition with native species, hybridization, predation, introduction of diseases, habitat disruption and trophic webs modification. The introduction of exotic species has been related to the extinction of 54 percent of aquatic native species worldwide. Overall, 70 percent of the extinctions of North American fishes and 60 percent of those from Mexico are related to non native species, totally or partially. Aquarium trade has shown an accelerated increase during the last decade with a trade value of US $160 million dollars. This increase parallels the boost of exotic species in the country. In fact, in the 80's only 55 non native fish species were registered in Mexico and by 2004 the number raised to 118, of which 67 (58.26 percent) have turned invasive. Several facts contribute to explain this: i) The low amount of varieties cultured in Mexico (61 varieties pertaining to 19 species), compared to the huge number of varieties imported (more than 700 from 117 families), ii) The number of fish imported in Mexico; 40 million ornamental fish are traded annually, of which 45 percent are imported (nearly 18 million fish were imported in 2006) while 55 percent are captive bred iii) there is a lack of official regulations for the establishment and operation of farms producing ornamental fish and for the translocation of ornamental fish within the country. As a result, ornamental fish species have been established in 9 out of 10 continental aquatic regions of Mexico. Some of these species have already severely impacted the environment and the economy in most regions of the country.

In: Invasive Species
Editors: Joaquín J. Blanco and Adrian T. Fernandes

ISBN: 978-1-61942-761-7

Chapter 1

Thermochemical Processing of Wood from Invasive Arboreal Species for Sustainable Bioenergy Generation and Activated Carbons Production

A. L. Cukierman[1,2,3*], G. V. Nunell[1,2], M. E. Fernández[1,2], J. De Celis[1], M. R. Kim[1], L. Gurevich Messina[1,2], and P. R. Bonelli[1,2]

[1]Programa de Investigación y Desarrollo de Fuentes Alternativas de Materias Primas y Energía (PINMATE) - Depto. de Industrias, Facultad de Ciencias Exactas y Naturales, Universidad de Buenos Aires. Int. Güiraldes, Ciudad Universitaria, Buenos Aires, Argentina

[2]Consejo Nacional de Investigaciones Científicas y Técnicas (CONICET), Buenos Aires, Argentina

[3]Cátedra de Farmacotecnia II. Depto. de Tecnología Farmacéutica, Facultad de Farmacia y Bioquímica, Universidad de Buenos Aires, Buenos Aires, Argentina

Abstract

Thermochemical conversion of wood sawdust from two invasive arboreal species (*Prosopis ruscifolia* and *Parkinsonia aculeata)* into valuable products is explored to provide novel possible means of spread control attempting to their use as alternative, easily renewable bio-resources for bioenergy generation and activated carbons production. Rapid pyrolysis of both wood species is examined to assess the feasibility of their conversion into energy products. Yields of the three kinds of pyrolysis products, comprising bio-oil, non-condensable gas, and bio-char, are determined from experiments performed in a bench-scale installation at 500 °C. The wood species noticeably affects yields and distributions of the pyrolysis products. Main physicochemical characteristics and higher heating value (HHV) of the products generated from *Prosopis ruscifolia*,

[*] E-mail: analea@di.fcen.uba.ar; anacuki@ffyb.uba.ar.

leading to higher bio-oil yield (~42%) than *Parkinsonia aculeata*, are determined. All the products show potentialities for energy applications. The bio-oil, further upgraded, could be employed as a liquid fuel with characteristics similar to those of fuel-oil, whereas the low heating value-gas evolved with pyrolysis course, mostly composed by CO_2, CO, CH_4 and H_2, might contribute to energy sustainability of the process. In turn, the bio-char (HHV≈ 27 MJ/kg) has potential as environmentally friendly solid bio-fuel. Kinetic measurements for the pyrolysis of the two wood species are also carried out by non-isothermal thermogravimetric analysis from room temperature up to 500 °C. An overall first-order decomposition model enables to properly represent kinetic data for both species over the whole temperature range. Differences in pyrolytic reactivity are reflected in the estimated kinetic parameters. On the other hand, use of wood sawdust from the two species as precursor of activated carbons targeted at wastewater treatment is investigated. For this purpose, the phosphoric acid activation process is applied at pre-established moderate conditions (acid/precursor weight ratio of 2, 450 ºC, 0.5 h). The activation atmosphere involved in the thermal treatment stage of the process is *in situ* modified to examine its influence on the development of porous structures, surface chemistry, and adsorptive behaviour of the resulting activated carbons. Either a self-generated atmosphere, flowing air or nitrogen is used. The activation atmosphere and the wood species have a marked influence on the characteristics of the derived carbons. Those developed from *Prosopis ruscifolia* wood in the self-generated atmosphere show higher BET surface area and total pore volume (2281 m^2/g, 1.7 cm^3/g) than the ones obtained under flowing air (1638 m^2/g and 1.3 cm^3/g), although the latter possess a higher total content of surface acidic/polar oxygen groups (2.2 meq/g) than the former (1.5 meq/g). Textural characteristics of *P. ruscifolia*-based activated carbons are superior to those for the carbons developed from *P. aculeata* wood in the self-generated atmosphere (968 m^2/g, 0.70 cm^3/g) and flowing N_2 (1103 m^2/g, 0.75 cm^3/g). These samples have total contents of surface acidic/polar oxygen functionalities of 1.9 meq/g and 1.7 meq/g, respectively. The ability of the activated carbons in removing representative toxic metal ions, oxo-anions, and organic compounds is examined from batch assays using dilute aqueous solutions as models of wastewater. Adsorption isotherms of Cu[II] ion onto *P. ruscifolia*-based activated carbons are determined and properly described by the Langmuir model. Maximum sorption capacity (X_m) for the air-derived carbons (X_m= 0.44 mmol/g) almost duplicates the value for those obtained in the self-generated atmosphere (X_m= 0.24 mmol/g), pointing to a predominant effect of the surface acidic functionalities on metal sequestering behaviour. Adsorption isotherms of salicylic acid onto the air-derived carbons are also ascertained, and satisfactorily fitted to the Langmuir model. On the other hand, effectiveness of the *P. aculeata*-based activated carbons in removing individual nitrate ion, phenol, and methylene blue at pre-established equilibrium conditions is comparatively investigated. High removal levels are attained for the organics (~ 90% and 100% for phenol and methylene blue, respectively), but nitrate removal is low (≤ 40%). Removal effectiveness of these carbons is inversely related to water solubility of the solutes and is slightly affected by the activation atmosphere. Accordingly, effective activated carbons of potential applicability in tertiary wastewater treatment may be developed from the two invasive species, as suitable, alternative precursors, by phosphoric acid activation at moderate conditions.

1. INTRODUCTION

The transfer of organisms among ecosystems and the spread of prolific species into new geographical areas cause growing concern on the repercussions in biodiversity conservation. Whether introduced by direct human movement of goods or by natural dispersal agents, some of these species become established beyond their native range, a subset of these spreads and some have negative impacts and are termed invasive (Keller et al., 2011). Successful invasion depends not only on the characteristics of the invading species but also on the characteristics, dynamics and history of the site that is invaded (Yates et al., 2004).

Woody plants were not widely recognized as invasive species of major importance until fairly recently, and currently 20% of the most intensively studied invasive species are of this kind. Some of the consequences arisen from plants invasions include the competition with and displacement of native species, the changes in plants community structure, the changes in the nutrient balance of soil and the loss of biomass and productivity of the site. Management efforts are underway to reduce economic and environmental problems associated with invasive trees. Different potential solutions have been proposed in order to control invasions (Yates et al., 2004; Richardson and Rejmánek, 2011). Among different strategies, thermochemical processing of biomass from invasive wood species to valuable products emerges as an attractive option because it is recognized as the most important available technology for biomass conversion.

Progressive increase in the demand for fuels, leading to a steep rise in oil price, together with exhaustion of fossil fuel reserves, pressure to reduce greenhouse gas emissions, and worldwide public concern about the adverse effects of environmental pollution related to the use of fossil fuels on human health and climate change, have deepened the interest in biomass utilization (González et al., 2008). Biomass is perceived to play a crucial role both as chemical feedstock and substitute for fossil fuels, due to its renewable character, minimal content of sulfur and nitrogen, and net carbon neutral energy conversion. Bioenergy offers cost-effective and sustainable opportunities with the potential to meet 50% of world energy demands during the next century and to attain the requirement of reducing carbon emissions from fossil fuels (Qu et al., 2011).

One promising route for thermochemical conversion of biomass is the pyrolysis process, resulting in the production of a huge number of chemicals. Pyrolysis is basically a polymeric structure cracking process in the absence of oxygen, enabling conversion of biomass into useful products. From the viewpoint of engineering applications, pyrolysis products are often lumped into three groups: permanent gases, a pyrolytic liquid (bio-oil/tar) and a carbon enriched solid product (char or bio-char). Several factors affect pyrolysis rate and yields as well as composition and properties of pyrolysis products. Temperature, pressure and heating rate are the chief operating parameters (Di Blasi, 2008). Generally, slow heating conditions, also termed carbonisation, favor char production, while high heating rates, known as fast and/or flash pyrolysis, enhance bio-oil yields. Although pyrolysis is one of the oldest technologies, where charcoal was originally the main product of interest, primarily for metallurgy, fast pyrolysis at moderate temperatures of around 500 °C and very short reaction times is currently considered as an emerging advanced technology for biomass conversion in terms of versatility, improved efficiency and environmental acceptability. It may be applied both as an integrated process for production of a liquid fuel that can be used directly and as an intermediate pretreatment step to convert solid biomass into a higher energy content transportable liquid for subsequent processing for heat, power, biofuels, and chemicals (Bozell, 2008; Bridgwater, 2011). Nevertheless, the development of a viable bio-renewable industry has to integrate all of the co-products of pyrolysis in the process of value adding (Schell et al., 2008; Yoder et al., 2011).

On the other hand, conversion of invasive wood species into activated carbons, whose development also involves thermochemical processes, constitutes another promising strategy which could contribute to control their spread. Activated carbons (ACs) are nowadays one of the most relevant adsorbent materials in several fields of human activity. They are used in a wide range of applications, in both gaseous and liquid phases, including medicinal uses, gas storage, pollutant and odour removal, gas separations, and catalysis due to their tunable textural properties (Silvestre-Albero et al., 2012). In particular, adsorption on granular activated carbons (ACs) is a proven, reliable technology for the advanced (tertiary) treatment of municipal and industrial wastewaters to remove small quantities of heavy metals and soluble organic compounds remaining in the wastewater following biological or physical-chemical treatments (US EPA, 2000; Rivera-Utrilla et al., 2011). ACs are widely used in decontamination processes because of their highly developed porous structures characterized by large specific surface areas. Activated carbons are mostly manufactured from coal, hardwood, coconut shells, and some polymers by thermal and chemical activation processes. The use of residues and sub-products from agriculture and industry as precursors for producing activated carbons has been intensively studied over the last decades (Rodriguez Reinoso, 2002; Bandosz, 2006; Bottani and Tascón, 2008; Demirbas, 2009; Girgis et al., 2011). However, the search for new, low-cost precursors, accessible and with potential to create significant economic valorisation is still needed in order to satisfy the increasing demand of activated carbons, especially for their use in water treatment applications (Basso and Cukierman, 2006; Nabais et al., 2008). Growth in consumption of activated carbons in current markets is forecast to be 5%py through 2015, with a potential total world demand to rise by nearly 10%py to 1.36Mt. Expansion of plants capacity for production of activated carbons as well as for water treatment is also foreseen (Roskill, 2008).

In this scenario, the present chapter deals with thermochemical conversion of wood sawdust from two invasive arboreal species (*Prosopis ruscifolia* and *Parkinsonia aculeata)* into valuable products in an attempt to provide novel possible means of spread control through the use of these species as alternative, easily renewable bio-resources for bioenergy generation and activated carbons production. *P. ruscifolia* is a species indigenous to south of South America, that grows very rapidly in Chaco region, the vast plain in Argentina, Paraguay, and Bolivia. It has the ability to fast invade degraded lands turning them unproductive (Cukierman, 2007; De Celis et al., 2009). In turn, *P. aculeata* is a major invasive species in several regions worldwide. It is native from America, and may be found in southwestern United States of America, Mexico, Bolivia, Ecuador, Galapagos Islands, Perú, Paraguay, Uruguay, and in different regions of Argentina. *P. aculeata* forms dense thickets, preventing access for humans, native animals and livestock to waterways. It grows easily on most types of soils, and withstands a great deal of heat, surviving in dry areas and for long prolonged dry spells. Moreover, its seeds remain viable for many years; the seedpods float, and the plant spreads by dropping pods into water, or pods are washed downstream by seasonal flooding (Cochard and Jackes, 2005).

Pyrolysis characteristics for the wood species are examined to assess the feasibility of their conversion into energy products. Yields of the three kinds of pyrolysis products, comprising bio-oil, gas, and bio-char, are determined from experiments performed in a bench-scale installation. Main physico-chemical characteristics and higher heating value for the wood species leading to the higher yield of bio-oil are determined by applying a set of complementary techniques. Kinetic characterization of the pyrolysis process is also carried out from measurements by non-isothermal thermogravimetric analysis from room temperature up to 500 °C, and modeling of the experimental results.

Furthermore, use of the sawdust of both wood species as precursor of activated carbons specifically targeted for wastewater treatment is investigated. For this purpose, phosphoric acid activation is applied at pre-established operating conditions by varying the activation atmosphere involved in the thermal treatment stage of the activation process in order to examine its influence on the development of porous structures and surface chemistry of the resulting activated carbons. Advantages of phosphoric acid activation in comparison with thermal or "physical" activation have been pointed out. They include a single thermal treatment stage, lower temperatures, and higher yields. In addition, most of the phosphoric acid can be recovered after the process is completed and its use as chemical reagent allows another degree of freedom for activated carbons tailoring. The process is considered particularly appropriate for conversion of lignocellulosic feedstocks into ACs (Rodríguez Reinoso et al., 2002; Basso and Cukierman, 2006). The ability of the developed activated carbons in removing representative toxic metal ions, oxo-anions, and organic compounds is determined from batch assays using dilute solutions as models of wastewater. It is interpreted in terms of the invasive species used as starting material, the resulting porous structures and surface chemistry which characterize the ACs, and the nature inherent to the pollutant examined.

2. EXPERIMENTAL

2.1. Materials

A piece of *Prosopis ruscifolia* wood without bark kindly provided by GESER (Facultad de Ciencias Exactas y Naturales, Universidad de Buenos Aires) was used for ACs preparation. *Parkinsonia aculeata* debarked trunks were obtained from a rural area located in Buenos Aires province, Argentina. They were sun dried, crushed, milled, and screen-sieved. The samples were then washed and dried at 60 °C up to constant weight. Proximate and elemental analyses of both invasive wood species were carried out by conventional ASTM standards and using a Carlo Erba EA 1108 instrument, respectively. Contents of the major biopolymeric constituents of the wood species, i.e holocellulose (celullose + hemicellulose) and lignin, and solvent extractive components were determined by applying the TAPPI standard methods. Higher heating value (HHV) was measured using a Parr 1341 oxygen bomb calorimeter. Sawdust samples from *Prosopis ruscifolia* and *Parkinsonia aculeata* are designated by PRS and PAS, respectively.

2.2. Pyrolysis Experiments

To obtain and characterize the three kinds of pyrolysis products, as well as to determine yields, experiments employing wood sawdust samples were carried out in a bench-scale installation. Basically, it consisted of a fixed-bed reactor (2.6 cm outer diameter, 110 cm total length) built in stainless steel. At the upper part of the reactor, a special device enabled to support a basket made of stainless steel mesh, used as container of the biomass samples. The bottom part of the reactor was externally heated by an electric furnace commanded by a temperature controller. The basket with the sample, constituting the solid fixed-bed, was centrally placed in the heated bottom zone of the reactor. In this way, the reduced length of the heated reactor zone enabled to attain short residence times of the volatiles evolved with the process course, thus minimizing possible extra-bed secondary reactions.

At the reactor outlet, a system for condensation and collection of the condensable volatiles generated with pyrolysis course was placed. It consisted of a series of four sequential flasks (250 cm^3), acting as cold tramps, immersed in a cooling bath. The first flask was empty, while the two latter were partially filled with acetone in order to ensure completely retention of the condensable vapors. Non-condensable vapors, after passing through the condensation system, were sampled periodically using Teflon gas bags for further analysis by gas chromatography. A Sep-Pak ®, Solid Phase Extraction C18 cartridge located before the main ventilation system was used to capture any residual liquid droplets formed downstream the condensation system. Finally, all the off-gases were sent to a main ventilation system.

All the experiments were carried out under isothermal conditions, based on the following procedure. First, the basket was loaded with the pre-weighed sample. A slip thermocouple was located inside the solid bed. Then, the loaded basket was introduced into the upper part of the reactor by means of the support device. Before heating the reactor, all the installation was purged from air by circulating a flow of N_2 (200 cm^3/min) for 1 h. Afterwards, the heating system was connected and the desired temperature was set. Once the pre-established temperature was attained, the basket containing the sample was displaced to the heated zone of the reactor. As determined from temperature measurements by the thermocouple inserted in the solid bed and monitoring the time required to attain the pre-established temperature, heating the sample in this way may be considered to take place almost instantaneously in all cases.

After the pre-established reaction time was achieved, heating was cut-off and the basket was immediately shifted towards the upper (non-heated) part of the reactor, keeping the N_2 stream. Once at ambient temperature, the basket was removed from the reactor. The residual solid and the accumulated liquid products contained in the flasks were weighted to determine product yields. These products were then carefully stored in closed containers for further characterization. Gas yields were obtained by difference from overall mass balances. Each experiment was repeated three times and averaged values are reported.

Preliminary assays were performed in order to examine the influence of the reaction time, sample mass (bed depth), particle diameter, and gas flow rate on process characteristics. Conditions used, for which variations were almost negligible, are reported in Table 1.

Table 1. Experimental conditions used for pyrolysis experiments

Temperature range	500 °C
Holding time	30 min
Total gas flow rate	200 cm^3/min
Sample mass	6 – 7 g
Particle diameter	0.5 – 1 mm

2.3. Characterization of Pyrolysis Products

2.3.1. Characterization of the Bio-char

Proximate analysis of the bio-char was carried out following conventional ASTM standards. Elemental composition (C, H, N, S) was assessed in a Carlo Erba EA 1108 instrument. Higher heating value (HHV) was measured using a Parr 1341 oxygen bomb calorimeter.

2.3.2. Characterization of the Liquid Products

Density of the liquid products collected at the end of pyrolysis experiments was determined by pycnometry at room temperature. To determine HHV and elemental composition, the liquid products resulting from each experiment were separated into two phases by addition of methylene chloride. In this way, an aqueous phase and an organic phase were obtained. HHV of the latter was determined after evaporation of the solvent under vacuum. For this purpose, the oxygen bomb calorimeter aforementioned was used following ASTM D240-02, for assessment of the heating value of liquid hydrocarbons. Elemental composition of the organic phase was also determined using the elemental analyzer earlier detailed.

2.3.3. Characterization of the Gaseous Products

Gas fractions evolved were analyzed by gas chromatography. A Shimadzu GC-8A instrument equipped with a thermal conductivity detector and an Altech CTR I concentric packed column was employed. Argon as carrier and a temperature of 25 °C were used. For quantification, calibration curves were previously assessed using a standard gas mixture SCOTTY® I Mix 237 (Scott Specialty Gases).

2.4. Kinetic Characterization of the Pyrolysis of Invasive Wood Samples

Kinetic characterization of the pyrolysis of the sawdust samples was performed by non-isothermal thermogravimetric (TG-DTG) analysis from ambient temperature up to 500 °C, operating under flowing nitrogen. A Netzsch STA 409 thermogravimetric balance equipped with a N_2 flow device and a data acquisition system was used to carry out kinetic measurements. Preliminary experiments were first conducted in order to assess operating conditions for which diffusional effects were negligible. Accordingly, masses of 10 mg, N_2 flow rates of 200 cm^3/min, and a heating rate of 10 °C/min were used. Further details of the experimental procedures employed have been depicted in previous own works (Bonelli et al., 2007; González et al., 2008).

2.5. Preparation of the Activated Carbons

Phosphoric acid activation was applied in order to develop activated carbons from both wood species, as precursors. Sawdust fractions (1 mm average particle diameter) were impregnated with a H_3PO_4 acid solution (50 wt%), using an acid/precursor weight ratio of 2.

The impregnated samples were first dried in an oven at 110 °C for 2 h. Afterward, they were placed in a horizontal, fixed-bed reactor externally heated by an electric furnace, and thermally treated at a heating rate of 3 °C/min up to 450 °C. Temperature selection was based on previous findings which demonstrate that maximum porosity development for ACs obtained by phosphoric acid activation from different lignocellulosic precursors is attained at 400-500 °C (Blanco Castro et al., 2000; Vernersson et al., 2002; Puziy et al., 2007). Thermal treatment of the impregnated sawdust samples was carried out under a self-generated atmosphere or a flow of air or nitrogen (0.1 m^3/h). Once the selected temperature was reached, it was held for 0.5 h. For experiments with flowing gases, either air or nitrogen flowed through the reactor throughout the thermal treatment stage and further cooling of the samples to reach ambient temperature. Further details of the equipment and procedure used have been earlier reported for other precursors (Blanco Castro et al., 2000; Basso et al., 2002; De Celis et al., 2009). To remove the excess of acid, the ACs were extensively rinsed with distilled hot water until neutral pH in the wash water was attained. Afterward, they were dried in an oven to constant weight. Yields were evaluated from weight differences.

The ACs developed from *Prosopis ruscifolia* wood under the self-generated atmosphere and flowing air are denoted as ACPR-S and ACPR-A, respectively. Likewise, those ACs derived from *Parkinsonia aculeata* wood under the self-generated atmosphere and flowing nitrogen are labeled as ACPA-S and ACPA-N, respectively.

2.6. Thermogravimetric Measurements of the Acid-impregnated Wood Sawdust Samples

In order to gain insight into the behaviour of the sawdust from the invasive wood species employed as precursors during the thermal treatment stage of the activation process, measurements by non-isothermal thermogravimetric analysis (TGA) were conducted for phosphoric acid impregnated samples. They were carried out on a Netzsch STA 409 thermal analyzer equipped with a data acquisition system. The impregnated samples were obtained by treatment with 50 wt% phosphoric acid solutions for a weight ratio of 2, prior to TGA analysis. The samples (~10 mg) were placed in the thermobalance, and heated from room temperature up to 500 ºC using a heating rate of 10 ºC min^{-1}.

2.7. Characterization of the Activated Carbons

2.7.1. Chemical Characterization

Chemical characteristics of the developed ACs were determined by proximate and/or elemental analyses. The former were performed according to American Society of Testing and Materials (ASTM) standards with the thermal analyser Netzsch STA 409. Elemental compositions of the samples were assessed using the Carlo Erba EA1108 elemental analyser.

The total and individual amounts of acidic/polar oxygen functional groups on the surface of the derived ACs were determined following a modified procedure based on Boehm's method (Basso et al., 2002). It enables to quantify carbonyls, phenols, lactones, and carboxyl groups by titration with a series of bases of different strength. Sodium ethoxide was used to determine the total amount of oxygen groups. Phenols, lactones, and carboxyls were quantitatively ascertained by titration with sodium hydroxide, whereas sodium bicarbonate allowed quantification of only carboxyl groups. Carbonyls were estimated by subtraction of NaOH-titrable groups from the total amount of acidic/polar oxygen functional groups, as determined by titration with sodium ethoxide. In turn, the quantity of phenols and lactones was evaluated by subtracting carboxyl groups from NaOH-titrable groups.

A total of 0.5 g of each sample was suspended in 50 mL of a 0.05 N solution of sodium ethoxide, sodium hydroxide, or sodium bicarbonate. The slurries were stirred for 24 h and afterward filtered. A 10 mL aliquot of the resulting solutions was added to 15 mL of a 0.05 N HCl acid solution. The solutions were subsequently back-titrated with 0.05 N NaOH. Values are expressed as miliequivalents per gram of sample. Likewise, the content of basic surface groups was determined by contacting 0.5 g of each activated carbon with 50 mL of 0.05 M HCl solutions. The slurries were stirred for 24 h and afterward filtered. A 10 mL aliquot of the resulting solutions was subsequently titrated with 0.05 N NaOH.

Besides, identification of surface functionalities of selected ACs was complementary conducted by Fourier Transform Infrared (FT-IR) spectroscopy. The spectra were recorded using a Nicolet Magna IR 550 spectrometer within the wavenumber range of 400-4000 cm^{-1}. Each sample was mixed with KBr and then ground in an agate mortar at an approximate ratio of 1:100 for preparation of the pellets. The resulting mixture was finally pressed. The background obtained from a scan of pure KBr was automatically subtracted from the sample spectra.

2.7.2. Textural and Morphological Characterization

N_2 adsorption-desorption isotherms at (-196 °C) for the derived ACs were determined with an automatic Micromeritics ASAP-2020 HV volumetric sorption analyzer. Prior to gas adsorption measurements, the samples were outgassed at 120°C overnight. Textural properties were assessed from the isotherms, according to conventional procedures depicted in detail in own previous studies (De Celis et al., 2009; Ramos et al., 2011). The Brunauer-Emmett-Teller (BET) surface area (S_{BET}) was determined by the standard BET procedure. Total pore volumes (V_t) were estimated from the amount of nitrogen adsorbed at the relative pressure of 0.95 ($p/p_0 = 0.95$). The mean pore radius (r) was calculated from $r = 4\ V_t / S_{BET}$.

2.8. Adsorption Assays

The ability of the prepared ACs to remove some representative pollutants from wastewater was investigated from batch adsorption experiments using model dilute solutions. The ACs developed from *P. ruscifolia* wood were tested for adsorption of Cu[II] ion and salicylic acid. Solutions of these species were used to mimic water polluted with toxic metals and metabolites, respectively. Besides, the ability in sequestering nitrate ion, phenol and methylene blue was examined for the ACs obtained from *P. aculeata* wood.

To investigate metal ions adsorption of the *P. ruscifolia*-based ACs, a stock solution (1000 mg/L) was first prepared by dissolving the necessary amount of analytical grade $CuCl_2.2\ H_2O$ in distilled water. Afterwards, it was diluted to obtain standard solutions of concentrations ranging between 5 and 400 mg/L.

Batch adsorption experiments using particles of 325 μm-average size were carried out at a constant temperature of 28 °C and pH = 6. The latter was selected on the basis of previous studies dealing with copper uptake by other developed ACs and/or commercial samples (Puziy et al., 2007). Concentrations of Cu[II] ion in solution were determined by UV-vis absorption spectrophotometry, according to a standard technique based on the formation of a brown compound of dithiocarbamate of Cu, followed by subsequent extraction of this compound with an organic solvent, and further analysis of the organic phase by spectrophotometry (De Celis et al., 2009).

The effect of the sample's dose on adsorption of Cu[II] ions was examined by contacting different weighed amounts of each sample (0.05–1 g) with 100 mL of 50 mg/L metal solutions under pre-established equilibrium conditions. Equilibrium was thoroughly verified for all the investigated systems from preliminary experiments performed for different prolonged contact times. Capped glass flasks containing the slurries were kept in a shaker at 500 rpm and controlled temperature (28 °C) for a period of 24 h. To check for any significant pH drift during the tests, the pH was recorded at the start and at the end of each experiment. No variations were determined with pH values maintained within 0.3 units. Once equilibrium was attained, the slurries were filtered through 0.45 μm membranes and metal concentrations in solution were determined as described above. The amount of Cu[II] ions sorbed at equilibrium per carbon mass unit (q_e) was calculated according to the following expression:

$$q_e = (C_0 - C_e) / D \quad (1)$$

where C_0, C_e denote the initial and equilibrium metal concentrations, respectively, and D, the dose of sample used.

Adsorption isotherms of Cu[II] ion were determined by contacting 0.4 g of each sample with 100 mL of solution of initial metal concentration in the range 5-400 mg/mL. The general procedure depicted above was followed, keeping the same experimental conditions. Likewise, adsorption isotherms of salicylic acid onto the ACs developed from *P. ruscifolia* wood under flowing air were assessed. In this case, doses of 0.5 g/100 mL, initial concentrations of salicylic acid in the range 10-200 ppm, and pH = 4.5 were used. Once equilibrium was attained, concentrations of salicylic acid in solution were determined by UV-vis absorption spectrophotometry at $\lambda = 297$ nm.

For the ACs developed from *Parkinsonia aculeata* wood under the self-generated atmosphere and flowing N_2, equilibrium adsorption of nitrate ion, phenol and methylene blue was investigated at pre-established conditions and optimal pH values. Equilibrium was thoroughly verified for all the investigated systems from experiments performed for different prolonged contact times. Samples' doses of 1 g/mL from solutions of 1.6 mM initial concentration of each solute were employed in these assays. As determined from preliminary measurements, pH values of 2 were used for nitrate ion and phenol tests, while pH = 7 was utilized for methylene blue assays. In all cases, concentrations of solutes in solution were determined by UV-vis absorption spectrophotometry (Shimadzu UVmini-1240 model) at $\lambda =$ 201 nm for nitrate ion, $\lambda = 270$ nm for phenol, and $\lambda = 663$ nm for methylene blue. The general procedure already depicted for adsorption experiments onto the *Prosopis ruscifolia* – based ACs was applied.

Duplicate experiments were at least conducted for every investigated system. Differences between replicates were less than 2% in all of the cases. Average values are reported. Furthermore, solute and adsorbent free blanks were used for control in all the tests.

3. Results and Discussion

3.1. Characteristics of Sawdust from the Wood Species

Table 2 reports results from proximate and elemental analyses for both wood species. Contents of the main biopolymers composing the species and higher heating value (HHV) are also included in the table.

Table 2. Proximate and elemental analyses of the sawdust from *Prosopis ruscifolia* (PRS) and *Parkinsonia aculeata* (PAS) wood

Proximate analysis (wt%, dry basis)	PRS	PAS
Volatile matter	75.4	77.4
Ash	2.4	1.4
Fixed carbon*	22.2	21.2
Elemental analysis (wt%, dry and ash free basis)		
C	47.1	48.0
H	6.2	5.7
N	0.4	0.2
O*	46.3	46.1
Biopolymers content (wt%, dry and extractive-free basis)		
Lignin	28	24
Holocellulose* (cellulose + hemicellulose)	72	76
HHV (MJ/kg)	18.9	18.7

* Estimated by difference.

Main differences in chemical characteristics lie in the contents of volatile matter, lignin and holocellulose, and in lesser extent in ash amount. As may be appreciated, HHV is similar for the two wood species and comparable to results reported for other ones (Saidur et al., 2011; Telmo and Lousada, 2011).

3.2. Yields and Properties of the Pyrolysis Products

Yields of the bio-char, bio-oil and gases generated from the rapid pyrolysis of the two wood species, on a dry-basis, are illustrated in Figure 1. The results show that the wood species noticeably affects yields and the distribution of the pyrolysis products. Pyrolysis of

PRS leads to a higher yield of bio-oil (42.3%) and similar yields of the bio-char and gases (~ 28-29%). Instead, similar yields of bio-oil and gases (~ 37-39%) arise from the pyrolysis of *Parkinsonia acueleata* wood. They are higher than the bio-char yield. Hence, pyrolysis of PAS yields overall greater quantities of fluid products than those produced from PRS. Bio-oil yields are rather lower than those reported for other wood species by using fluidized bed reactors (Butler et al., 2011).

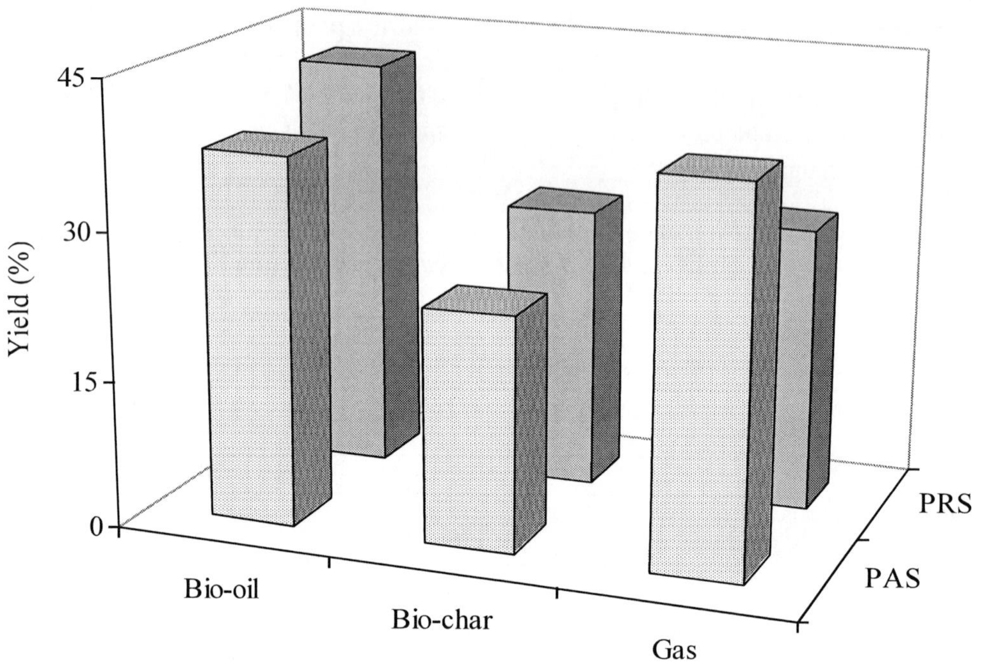

Figure 1. Yields of the solid, liquid and gaseous products generated from the pyrolysis of *Prosopis ruscifolia* (PRS) and *Parkinsonia aculeata* (PAS) wood sawdust at 500 °C.

Formation and yield of the bio-char have been ascribed to dehydration, slow decarboxylation, depolymerization, and recombination of the decomposition products taking place during pyrolysis of biomass at relatively low temperatures, as the main responsible reactions (White et al., 2011). Contents of bio-polymers composing the wood samples also affect yield. The higher yield of the PRS – derived bio-char compared to that resulting from PAS is consistent with the greater content of lignin of the PRS wood, likely because lignin is more resistant to degrade than cellulose and hemicellulose due to its aromatic structure. Lignin is a natural amorphous polymer consisting of phenylpropane units, and their precursors are three aromatic alcohols, namely, p-coumaryl, coniferyl, and sinapyl alcohols (Basso et al., 2005). In turn, the higher yield of gaseous products attained in the pyrolysis of PAS is in line with the higher content of volatile matter characterizing this species. Tables 3 and 4 show physicochemical properties of the bio-char and bio-oil resulting from pyrolysis of *Prosopis ruscifolia* wood, respectively.

Table 3. Physicochemical properties of the bio-char obtained from pyrolysis of *P. ruscifolia* wood

Proximate analysis (wt%, dry basis)	
Volatile matter	10.6
Ash	6.7
Fixed carbon*	82.7
Ultimate analysis (wt%, dry and ash free basis)	
C	81.4
H	2.7
N	0.8
O*	15.1
HHV (MJ/kg)	27.2

* Estimated by difference.

Comparison of the data shown in Tables 2 and 3 indicates that release of volatile matter upon pyrolysis of PRS sawdust results in a bio-char with enhanced contents of ash and fixed carbon, in agreement with reported general trends (Della Rocca et al., 1999; González et al., 2008; Özçimen and Ersoy-Meriçboyu, 2010). Pyrolysis also promotes enrichment in %C accompanied by reductions in %H and %O, arising from release of water vapor, carbon oxides, hydrogen, and volatile hydrocarbons, thus favoring the potential of PRS wood as solid bio-fuel. These changes lead to a significant increase in HHV, as evidenced by the values determined for the pristine PRS sawdust and the derived bio-char, indicating that it has potential as solid bio-fuel. HHV of the latter is within the range characterizing commercial wood charcoal as well as those determined for bio-char from several wood species and other lignocellulosic biomasses (Della Rocca et al., 1999; Cordero et al., 2001; Bonelli et al., 2003). Besides, HHV experimentally determined for the raw wood and derived bio-char are in close agreement with values estimated by a multivariate unified correlation proposed in the literature, based on the elemental composition and ash content of the samples (Channiwala and Parikh, 2002).

According to international standards (Encinar et al., 1997; Basso et al., 2005; González et al., 2005), fixed carbon contents higher than 76 wt% are required for charcoal to be used in the manufacture of briquettes. Therefore, on the basis of this requirement, the derived bio-char has potential for this purpose. Another criterion adopted as an index of good quality takes into account volatile matter content of the char together with the intended application. For domestic use, volatile matter contents ranging between 20% and 30% are acceptable, whereas these contents are restricted to 10-15% for use in the steel industry. Since the content of volatile matter for the *Prosopis ruscifolia*–derived bio-char is around 11%, it is potentially suited for both applications.

With respect to the liquid product, bio-oil is reportedly composed of a very complex mixture of oxygenated hydrocarbons with an appreciable proportion of water from both the original moisture and reaction product. It can be considered a micro-emulsion in which the continuous phase is an aqueous solution of holocellulose decomposition products, that stabilises the discontinuous phase of pyrolytic lignin macro-molecules through mechanisms such as hydrogen bonding (Bridgwater, 2011). The heating value and elemental composition of the *P. ruscifolia* derived bio-oil (Table 4) are comparable to those reported for other ones (García Perez et al., 2008). Compared to crude oil, a crucial difference is the higher oxygen content (37%) of the derived bio-oil, reportedly affecting homogeneity, polarity, heating value, and acidity. The present bio-oil shows a homogenous appearance, likely because the oxygenated molecules of low molecular weight, especially alcohols and aldehydes, could act as surfactant for the higher molecular weight compounds, which are normally considered non-polar and immiscible with water. The higher contents of water and oxygen are considered responsible for the lower heating value characterizing bio-oils, which is almost half that of crude oil (Mortensen et al., 2011).

Table 4. Characteristics of the bio-oil obtained from pyrolysis of *P. ruscifolia* wood

Ultimate analysis* (wt%, dry and ash free basis)	
C	54.4
H	8.4
N	0.6
O**	36.6
Density (g/cm^3)	1.04
pH	2.3
HHV* (MJ/kg)	25.1

* Corresponding to the organic fraction.

** Estimated by difference.

The pH and density of the *P. ruscifolia*-derived bio-oil (Table 4) are also within the range of reported values. The acidic nature of the bio-oils, with pH values between 2 and 4, is associated with contents of acetic acid and formic acid, and poses corrosion problems to equipment used for storage, transport, and processing. Besides, bio-oil density is higher than that for crude oil, also attributable to its high oxygen content. Main unfavorable characteristics affecting bio-oil quality have been associated with oxygenated compounds, such as carboxylic acids, ketones, and aldehydes. In addition, it should be mentioned that an overall decrease in bio-oil quality usually takes place during storage, ultimately resulting in phase separation (Gupta and Demirbas, 2010; Mortensen et al., 2011). Accordingly, utilisation of bio-oil requires a general decrease in the oxygen content in order to separate the organic phase from water as well as to increase the energy content and chemical stability. Upgrading of bio-oils has been extensively reviewed in the literature. Bio-oil can be upgraded in a number of ways, physically, chemically or catalytically, to attain a storable and transportable liquid and/or to supply valuable chemicals. Although demonstration of upgrading technologies is underway, it could take some time before they are deployed on an industrial scale (Bridgwater, 2011; Butler et al., 2011). Seemingly, the simplest way to use bio-oil as a transport fuel is to combine it with diesel directly through emulsification by the aid of a surfactant (Zhang et al., 2007).

On the other hand, main gaseous species evolved with pyrolysis course comprise CO_2, CO, CH_4, H_2 and, to a lesser extent, C_2H_6, and C_2H_4. Typical concentration-time profiles for the main species are exemplified in Figure 2, for the pyrolysis of *Prosopis ruscifolia* wood. Figure 3 shows the amount of moles produced per mass unit of biomass (G_i).

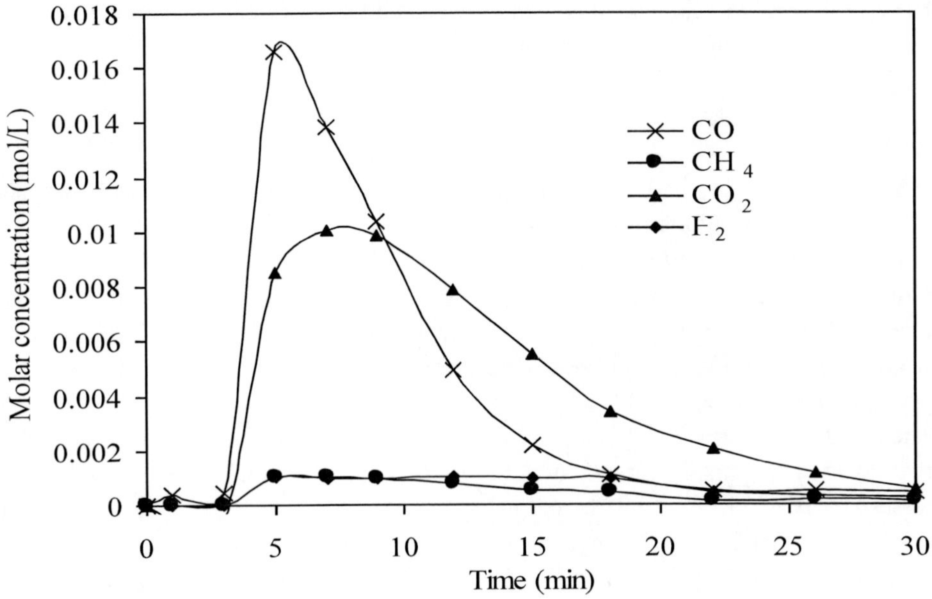

Figure 2. Time-concentration profiles of the main gaseous species generated in *P. ruscifolia* wood pyrolysis at 500 °C.

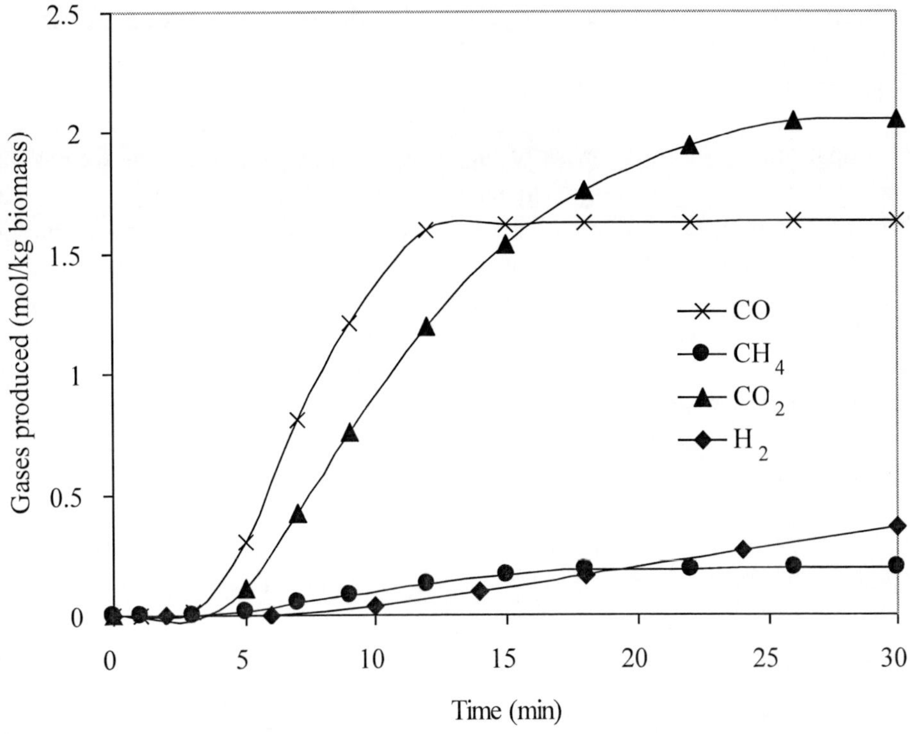

Figure 3. Amount of moles produced of each gaseous species per mass unit of biomass as a function of time for *P. ruscifolia* wood pyrolysis at 500 °C.

HHV of the gas was calculated considering G_i and the heat of combustion for each of the gaseous species, according to the following equation (2):

$$HHV\ [MJ/kg] = 0.802\ G_{CH4} + 0.286\ G_{H2} + 0.283\ G_{CO} \quad (2)$$

G_{CH4}, G_{H2}, and G_{CO} in Eq. (2) are the total amount of each of the gaseous species produced per mass unit of biomass.

For the pyrolysis conditions examined, HHV of the gas product is 0.8 MJ/kg of biomass (8 MJ/m^3). It agrees with some values reported in the literature for pyrolytic gases arising from other biomassic feedstocks for similar conditions (Encinar et al., 2009). The gas produced could contribute to provide heat, which may help to the sustainability of the pyrolysis process (Bridgwater, 2011).

The results in Figures 2 and 3 show that CO_2 is the main gaseous species arising from *P. ruscifolia* wood pyrolysis, in agreement with results reported for other woody and agricultural biomasses (Di Blasi et al., 1999; Ioanidou et al., 2009; Qu et al., 2011). According to the literature, CO_2 is mainly produced from reforming and cracking of carbonyl and carboxyl functional groups. In turn, CO is generated when C−O−C and C=O are broken, while CH_4 is produced from decomposition of methoxyl−O−CH_3. H_2 mainly arises from cracking and deformation of C=C and C–H groups (Park et al., 2011; Qu et al., 2011).

3.3. Kinetic Characterization of Pyrolysis of the Invasive Wood Species

Typical non-isothermal TG curves for the pyrolysis of *P. ruscifolia* and *P. aculeata* wood sawdust are illustrated in Figures 4 and 5, respectively. In these figures, instantaneous weight losses normalized by the initial mass of the sample, namely instantaneous weight fractions (w), are represented as a function of the temperature (T). The reaction rate profiles obtained by differentiation of the normalized weight losses-time curves are also included in the figures. Main characteristics of the TG curves and reaction rate profiles are summarized in Table 5.

The results (Figures 4-5, Table 5) indicate that pyrolysis of PRS and PAS starts at 175 °C and 200 °C, respectively, releasing volatile matter. TG curves show two major weight loss stages. For PRS, the first stage takes place in the range 200-360 °C and the second one, above 360 °C. In turn, these stages occur between 250 °C and ~ 380 °C, and above the latter temperature for PAS. In both cases, the slope of the curves changes between the two temperature intervals, with steeper slopes for the first stage. It may also be seen that the first stage proceeds more abruptly for PAS than for PRS. Besides, reaction rate profile for the pyrolysis of PAS is shifted to higher temperatures and a higher maximum reaction rate (0.067 min^{-1}, 350 °C) is attained compared to that of PRS (0.054 min^{-1}, 321 °C).

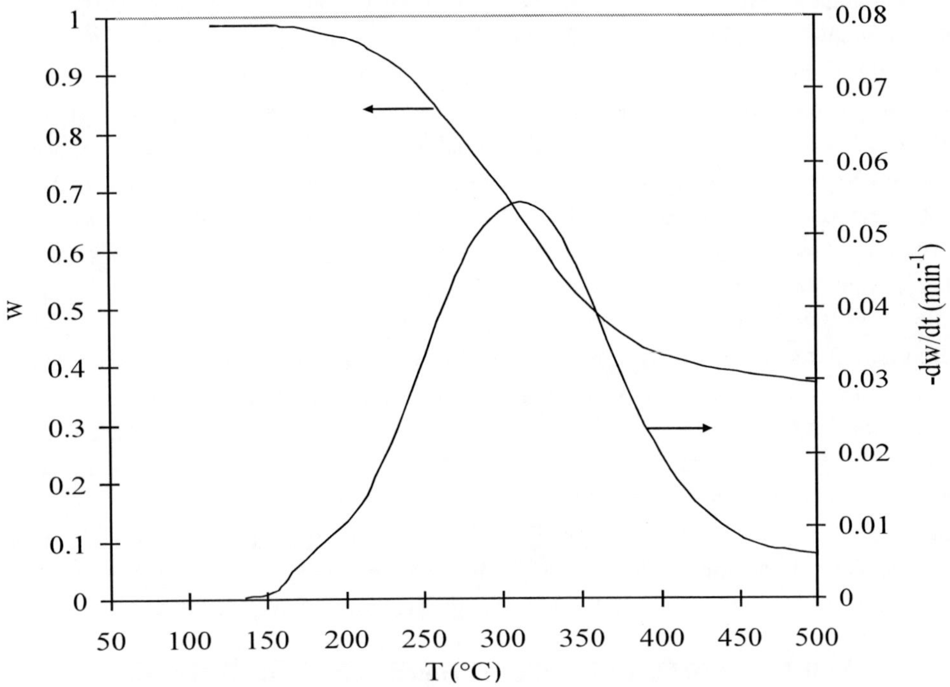

Figure 4. TG-DTG curves for the pyrolysis of *P. ruscifolia* wood sawdust.

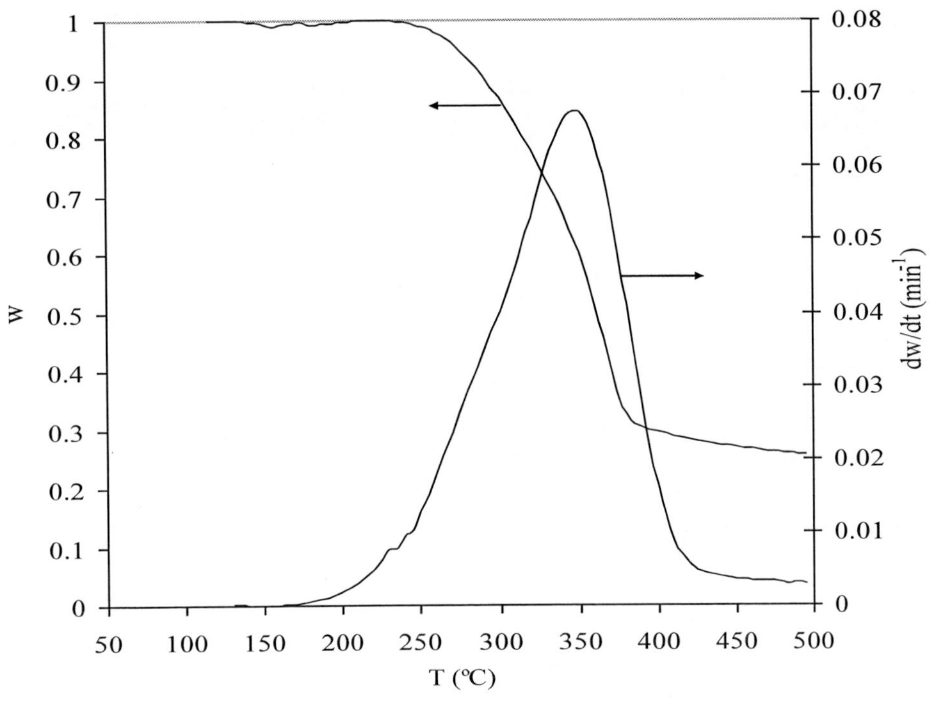

Figure 5. TG-DTG curves for the pyrolysis of *P. aculeata* wood sawdust.

In agreement with other reported results, distinct weight loss zones may be associated with degradation dynamics of main biopolymers constituting wood, namely cellulose, hemicellulose, and lignin. Decrease in weight fractions at around 225–325 °C is generally attributed to the onset of hemicellulose degradation, while contribution of cellulose decomposition to pyrolysis is considered to predominate in the range between 325 and 375 °C. Lignin has a broad decomposition temperature range at temperatures higher than 250 °C. Hence, degradation of cellulose and hemicellulose, possessing a polysaccharide structure relatively easy to breakdown, should take predominantly place at the lower temperatures. At higher temperatures, decomposition of lignin, which is more resistant to degrade than cellulose and hemicellulose, due to its cross-linked aromatic structure, appears to become predominant inducing pyrolysis of biomass to slow down (Basso et al., 2005; González et al., 2008; Wongsiriamnuay and Tippayawong, 2010). Process simulations have also shown that hemicellulose and cellulose decompose independently of one another, whereas lignin degrades slowly over a broad range of temperatures (Di Blasi, 2008). The residual weight fraction at the highest temperature of the investigated range (500 °C) is higher for the pyrolysis of PRS than that attained for PAS (Table 5). The trend agrees with the higher bio-char yield determined for the former in the fixed-bed reactor experiments conducted at the same temperature (Figure 1), attributable to the greater lignin content of PRS.

Table 5. Characteristics of the TG curves and reaction rate profiles for the pyrolysis of pristine and phosphoric acid pre-treated wood sawdust from *Prosopis ruscifolia* (PRS) and *Parkinsonia aculeata* (PAS) species

Sample / Characteristic	PRS	PAS	Acid treated PRS	Acid treated PAS
T_{onset} °C	175	200	100	107
T_{peak} °C	321	350	184	183
T_{end} °C	440	420	350	340
$(dw/dt)_{peak}$ (min^{-1})	0.054	0.067	0.0134	0.0096
w_{peak}	0.65	0.60	0.79	0.88
$w_{T=500\,°C}$	0.37	0.25	0.54	0.66

To evaluate the kinetic parameters for the pyrolysis of both wood species, modeling of the TG curves was carried out. The simple model which assumes pyrolysis as a first-order overall decomposition and the usual Arrhenius type dependence of the rate constant on temperature, was applied:

$$-dw/dt = k_0 \exp(-E_A/RT)(w - w_r) \quad (3)$$

w and w_r in Eq. (3) are the instantaneous and residual weight fractions, respectively; k_0 is the pre-exponential factor, E_A, the activation energy, R, the universal gas constant, and T, the absolute temperature. Model characteristic parameters (k_0, E_A) were evaluated by non-linear regression analysis. They are reported in Table 6 along with standard deviations (s).

Model predictions for the pyrolysis of *Prosopis ruscifolia* and *Parkinsonia aculeata* wood sawdust are shown in Figures 6 and 7, respectively. As seen, the model succeeds in representing properly the experimental data for both wood species over the whole range of temperatures, with reasonable s values. Values of the activation energy are quite similar, but k_0 parameter differs substantially, reflecting differences in the pyrolytic reactivity between the wood species.

Table 6. Kinetic parameters estimated for the pyrolysis of *Prosopis ruscifolia* (PRS) and *Parkinsonia aculeata* (PAS) wood sawdust

Sample / Parameter	k_0 (min^{-1})	E_A (kJ/mol)	s (%)
PRS	2.9 x 10^5	56.9	2
PAS	5.0 x 10^3	55.6	4

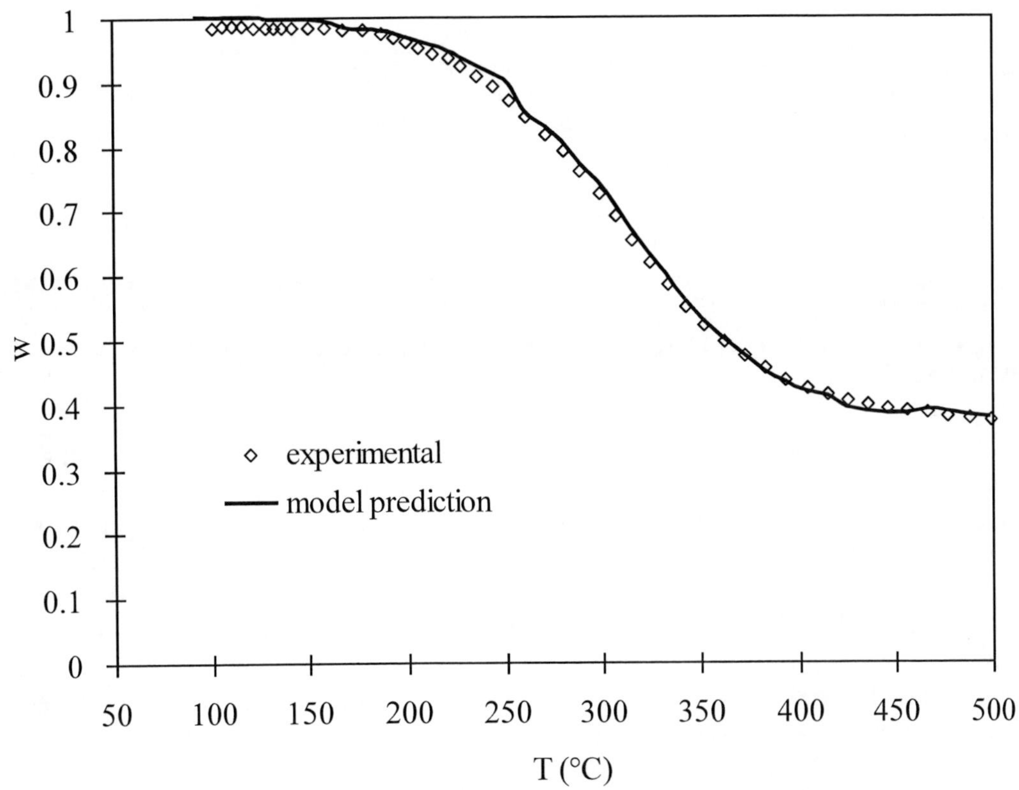

Figure 6. Comparison between experimental data and model predictions for the pyrolysis of *Prosopis ruscifolia* wood.

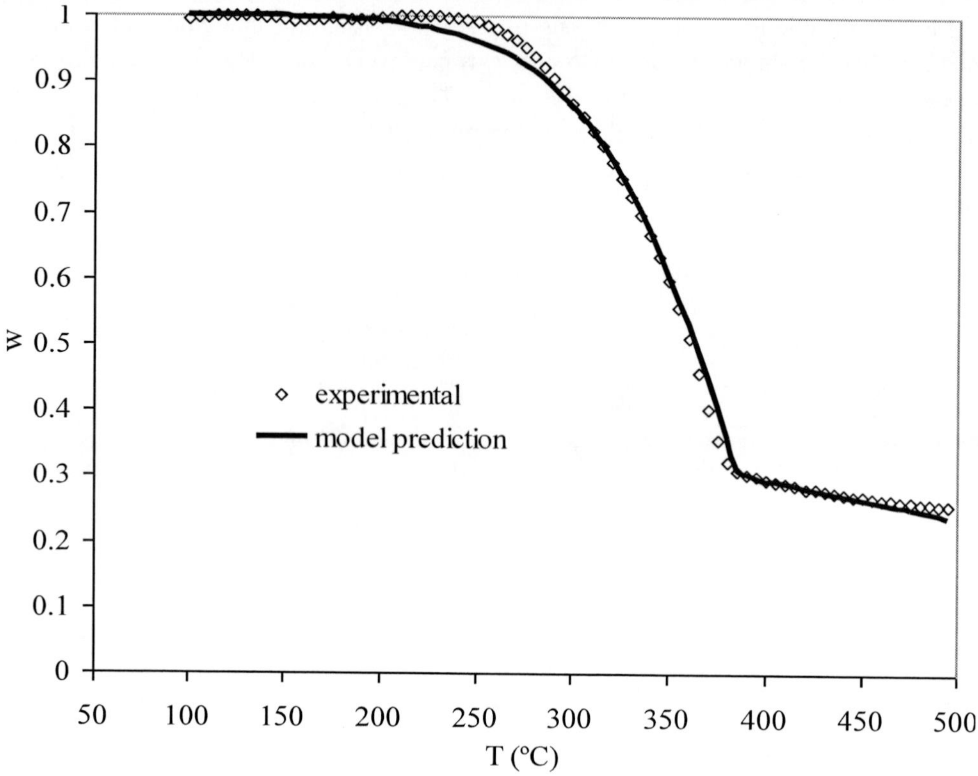

Figure 7. Comparison between experimental data and model predictions for the pyrolysis of *Parkinsonia aculeata* wood.

3.4. Yield and Thermal Degradation Behaviour of the Acid-pre-treated Wood Sawdust

An important aspect of the activation process is the amount of precursor which is converted into final product (Nabais et al., 2010). Overall yields for the activated carbons developed from both wood precursors under the different atmospheres are shown in Figure 8. As may be observed, phosphoric activation of *P. aculeata* under flowing nitrogen leads to the highest yield. The results for the activated carbons obtained from the two precursors under the self-generated atmosphere (ACPR-S and ACPA-S) attest the influence of the wood species used as precursor.

Typical TG curves and pyrolysis rate profiles for the phosphoric acid-impregnated *Prosopis ruscifolia* and *Parkinsonia aculeata* wood samples are illustrated in Figures 9 and 10, respectively. Weight fractions (w) and pyrolysis rates (-dw/dt) are shown as a function of temperature in the figures. As for the untreated wood samples, pyrolysis rate profiles were obtained by differentiation of weight fractions vs time curves for each sample. Main characteristics of the TG curves and reaction rate profiles are listed in Table 5.

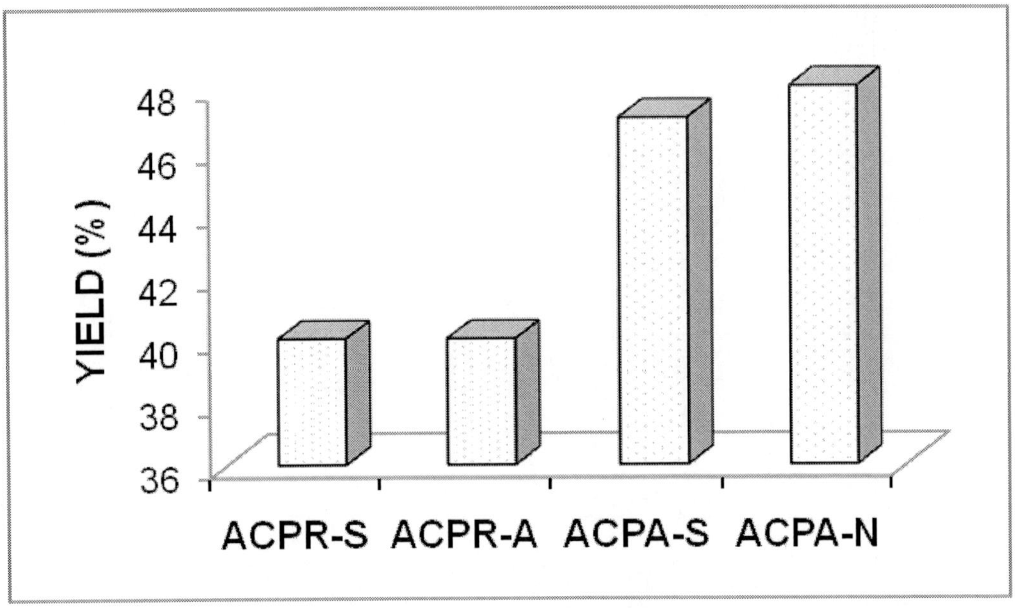

Figure 8. Overall yields of the activated carbons developed from from *P. ruscifolia* and *P. aculeata* wood under the self-generated atmospheres (ACPR-S, ACPA-S), flowing air (ACPR-A) or nitrogen (ACPA-N).

The results in Figures 9 and 10 show that pyrolysis of the acid-impregnated samples presents a noticeably different thermal degradation behaviour compared to those determined for both pristine wood species (Figures 4 and 5). As seen, the acid-treated samples start to decompose at lower temperatures (100-107 ºC) than the pristine ones, degradation proceeding steadily throughout the whole range of temperatures, including the low temperature region where no weight losses for the untreated species were detected. Phosphoric acid appears to facilitate the reactions in the wood samples that lead to weight loss at the initial stages of carbonization. Likewise, while maximum pyrolysis rates for the untreated PRS and PAS occur at 321 ºC and 350 ºC, respectively, they are shifted to lower temperatures (183-184 ºC) and attain lower values for the acid-treated samples (Table 5). At 500 ºC, higher values of the residual weight fractions characterize the pyrolysis of the acid-treated species in comparison with those of the untreated ones.

The behaviour is similar to that reported for the thermal degradation of other phosphoric acid-treated lignocellulosic biomasses (Girgis and Ishak, 1999; Vernersson et al., 2002; Basso et al., 2005). It may be attributed to modifications in physicochemical properties of the wood samples induced by the phosphoric acid, affecting markedly its thermal behaviour. Seemingly, the acid acts as a catalyst promoting bond-cleavage reactions that favor conversion of aliphatic to aromatic compounds at lower temperatures than for wood degradation in the absence of the reagent (Jagtoyen and Derbyshire, 1998; Benaddi et al., 1998). In turn, alterations in the samples' structure caused by incorporation of the acid also appear to favor cross-linking reactions, decreasing losses of volatile compounds and leading to higher weight fractions for the acid-treated samples (Blanco Castro et al., 2000; Puziy et al., 2002; Ramos et al., 2007).

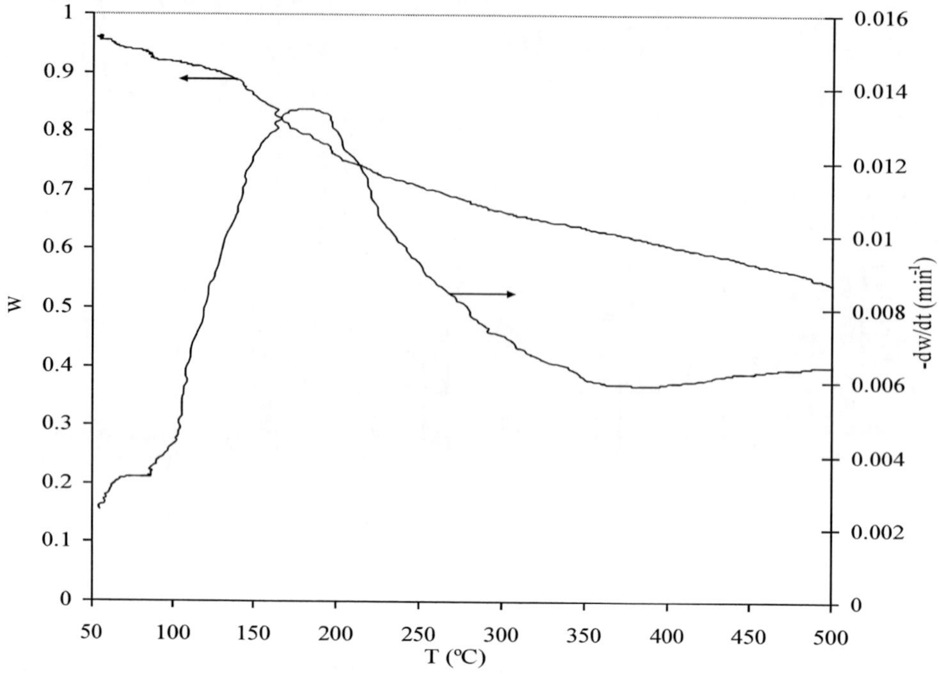

Figure 9. TG-DTG curves for the pyrolysis of phosphoric acid pre-treated *Prosopis ruscifolia* wood sawdust.

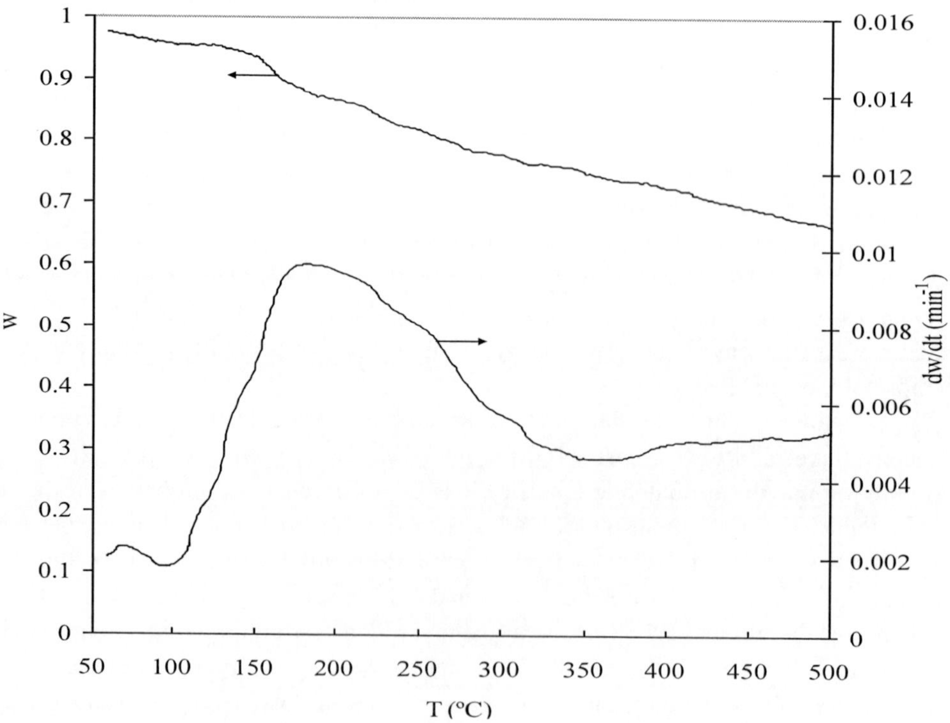

Figure 10. TG-DTG curves for the pyrolysis of phosphoric acid pre-treated *Parkinsonia aculeata* wood sawdust.

3.5. Chemical Characteristics of the Developed Activated Carbons

Chemical characteristics of the activated carbons derived from both wood species under the different atmospheres are summarized in Table 7. As inferred from comparison of the results in Tables 2 and 7, noticeable changes in the precursors occur due to the activation process, in turn depending on the activation atmosphere employed in the thermal treatment stage. Likewise, comparison of the results for the ACs obtained from the two species in the self-generated atmospheres for the same experimental conditions point to the effect of the wood species used as precursor.

Table 7. Chemical characteristics of the activated carbons developed from *P. ruscifolia* and *P. aculeata* wood under the self- generated atmospheres (ACPR-S, ACPA-S), and under flowing air (ACPR-A) or nitrogen (ACPA-N)

Sample	ACPR-S	ACPR-A	ACPA-S	ACPA-N
Ash (%)	3.0	8.0	8.3	13.6
%C	88.0	81.1	77.0	82.3
%H	2.4	2.1	2.5	2.5
%N	0.2	0.3	0.3	0.4
%O*	9.4	16.5	20.2	14.8

* Estimated by difference.

Contents of ash and elemental oxygen of the ACPR-A are markedly higher than those of the ACPR-S, in agreement with the stronger oxidation conditions used in the preparation of the former. The trend also holds for the oxygen content of the ACs derived from *P. aculeata* developed in the self-generated atmosphere (APA-S) and under flowing nitrogen (ACPA-N), with the former possessing a higher content than the latter. However, the ACs developed from *P. aculeata* wood under flowing N_2 exhibit a larger content of ash than the samples resulting from both species in mild and strong oxidation conditions. It could be attributed to the formation of phosphor compounds poorly soluble in water that seems to remain in the carbon after washing when carbonization of acid treated samples is carried out in an oxygen-free atmosphere (Vernersson et al., 2002).

The surface of ACs has been described as a collection of organic functional groups containing oxygen, with these groups occurring primarily at the edges of broken graphitic planes, and basal planes that consist of large fused aromatic ring systems in a graphite-like structure (Patrick, 1995; Rodríguez Reinoso, 2002). The total and individual contents of surface acidic/polar oxygen functional groups of the ACs derived from *P. ruscifolia* and *P. aculeata* in the different atmospheres are illustrated in Figure 11.

As observed in Figure 11, *in situ* modification of the activation atmosphere pronouncedly affects the development of acidic/polar oxygen functionalities on the surface of the carbons obtained from both wood species, in agreement with results reported for ACs derived from other starting materials (Basso et al., 2002; Basso and Cukierman, 2005, 2006; Puziy et al., 2007). The ACPR-A obtained under severe oxidation conditions exhibits a higher total content of acidic oxygen functional groups in comparison with the ACs developed with restricted supply of oxygen, as that arising from the self-generated atmospheres or under flowing nitrogen. In general, the trend is consistent with the elemental oxygen content of the samples. The presence of oxygen in the activation atmosphere seems to favour a preferential development of phenols, lactones, and carbonyls for both wood species, as inferred from the major proportions of these functionalities on the surface of ACPR-S, ACPR-A, and ACPA-S. Seemingly, these groups are formed through the reaction between the precursor and oxygen present during activation. Instead, the conditions generated by operating under N_2 flow lead to greater quantities of carbonyls and carboxylic groups, despite the susceptibility of the latter to decompose at relatively low temperatures. Contents of carboxylic groups of the ACs derived from *P. aculeata* in both atmospheres are higher than those of the *P. ruscifolia*-based samples. These functionalities likely arise from acidic hydrolysis of the raw material used as a precursor (Guo and Rockstraw, 2007). Likewise, the ACs developed from *P. aculeata* in both atmospheres possess 0.1 mequiv/g of basic functional groups. Those based on *P. ruscifolia* under flowing air have 0.4 mequiv/g, but no basic groups were detected for the ACs obtained in the self-generated atmosphere. Accordingly, the results indicate that, for each species, *in situ* modification of the activation atmosphere not only influenced the total amount of acidic/polar surface oxygen groups, but also the relative proportions of the individual functionalities present on the surface of the resulting ACs.

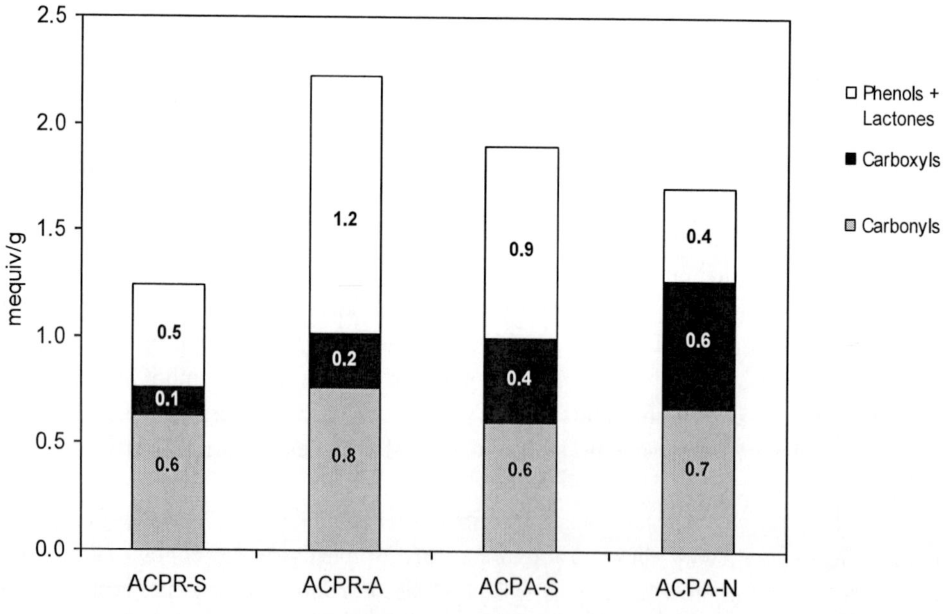

Figure 11. Total and individual quantities of acidic/polar surface functional groups determined for the activated carbons developed from *Prosopis ruscifolia and Parkinsonia Aculeata* wood sawdust under the different atmospheres.

In order to gain insight into the effect of modifying the activation atmosphere on the developed surface functionalities, illustrative examples of FT-IR spectra for the ACs derived from *P. ruscifolia* wood sawdust in the self-generated atmosphere and under flowing air are shown in Figure 12.

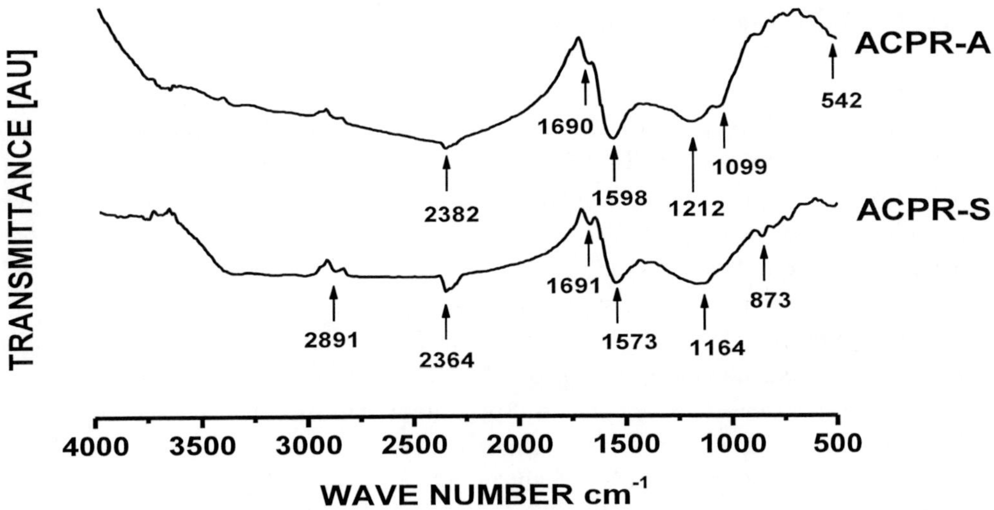

Figure 12. FT-IR spectra for the activated carbons developed from *P. ruscifolia* wood under the self-generated atmosphere (ACPR-S) and flowing air (ACPR-A).

The spectra for the ACs developed under both atmospheres show absorption bands between 500 and 850 cm^{-1}, that may be assigned to aromatics substituted by aliphatic groups. They also display a broad band between 950 and 1300 cm^{-1}, frequently found in ACs obtained by phosphoric acid activation. Presumably, absorption in this region arises from superposition of signals corresponding to oxygen functional groups, like ethers, phenols and lactones, and also to phosphorus and phosphor-carbonaceous compounds present in phosphoric acid activated carbons. Formation of the latter, which are thermally stable even at high activation temperatures, has been attributed to the reaction between the precursor or its acidic hydrolysis products and H_3PO_4 acid or other forms of phosphorus-containing acids, arising from pyro- or polyphosphoric acids formed at the activation temperature. The broad bands at 1000-1300 cm^{-1} have been assigned to C–O single bonds such as those in ethers, phenols, acids and esters (Puziy et al., 2002, 2007; Guo and Rockstraw, 2007).

In addition, the band between 1400 and 1750 cm^{-1} is preferentially assigned to carboxyl groups, quinones, ketones, lactones, diketone and keto-esther, and keto-enol. The latter band is more pronounced for ACPR-A, whereas the relation of intensities is slightly opposite for the ACPR-S. Both samples also present an intense band at 1550 and 1650 cm^{-1} resulting from C=C stretching vibrations in aromatic rings enhanced by polar functional groups. The presence of a band at around 1690 cm^{-1} is also observed for both *Prosopis* wood-based ACs. It may be due to the streching vibrations of C=O moieties in carboxylic acid groups, esters, lactones and quinones. Likewise ACPR-S and ACPR-A show absorption bands at 2300 and 2900 cm^{-1}. The peaks at 2380 cm^{-1} for the ACPR-A, and at 2364 cm^{-1} for the ACPR-S, may be assigned to carbon-oxygen bonds in ketene groups. The peak at 2890 cm^{-1} could be originated from C–H stretching in $-CH_2-$ and $-CH_3$ groups. Furthermore, the OH bond corresponding to alcohols, phenols and water chemisorbed is in the range between 3600 and 3200 cm^{-1} (Puziy et al., 2002, 2007; Guo and Rockstraw, 2007; Fierro et al., 2007). In summary, main differences in the spectra of the *Proposis ruscifolia* wood-based ACs are found in the region comprised between 500-1700 cm^{-1}, that could affect their further behaviour in adsorption.

3.6. Textural Characteristics of the Developed Activated Carbons

Figures 13 and 14 display N_2 adsorption-desorption isotherms for the ACs developed from *Prosopis ruscifolia* and *Parkinosea aculeata* wood, respectively. Textural characteristics evaluated from the isotherms are listed in Table 8. The shape of the adsorption isotherms for the ACs developed from *Prosopis ruscifolia* under the self-generated atmosphere and flowing air as well as those for the *Parkinsonia aculeata*-based ACs is quite similar, and shows characteristics between those of Type I and II, according to IUPAC classification. They point to the presence of micropores (< 2 nm) and mesopores (2-50 nm).

According to reported mechanisms (Jagtoyen and Derbyshire, 1998; Blanco Castro et al., 2000), dilation and stabilizing formation of cross-links in the form of phosphate esters are responsible for the extensive development of porosity within ACs. It has been proposed that bond-cleavage reactions take place at low temperatures, in the range between 50 and 200 °C, yielding fragments of the precursor's constituent biopolymers. At higher temperatures (200–450 °C), these fragments combine with H_3PO_4 acid forming phosphate linkages which give rise to a rigid, cross-linked structure. The structure dilates and cross-linking reactions dominate over bond-cleavage reactions, giving rise to an expanded porous matrix after acid removal.

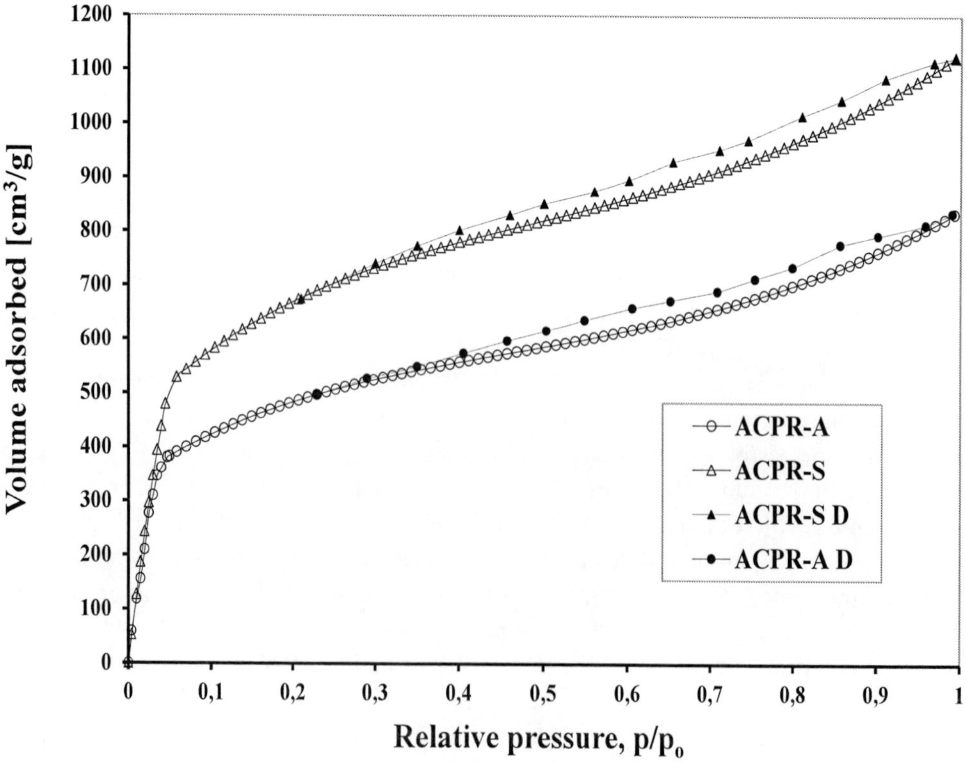

Figure 13. N_2 adsorption (empty symbols) and desorption (solid symbols) isotherms for the activated carbons developed from *Prosopis ruscifolia* wood sawdust under the self-generated atmosphere (ACPR-S) and flowing air (ACPR-A).

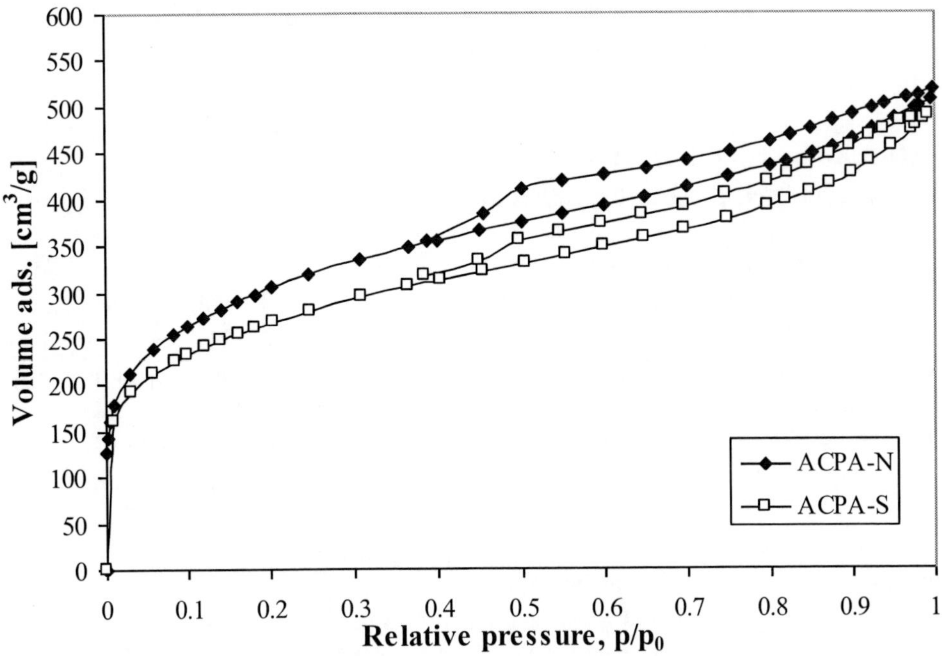

Figure 14. N_2 adsorption - desorption isotherms for the activated carbons developed from *Parkinsonia aculeata* wood sawdust under the self-generated atmosphere (ACPA-S) and flowing N_2 (ACPR-N).

Table 8. Textural properties of the activated carbons developed from *Prosopis ruscifolia* and *Parkinsonia aculeata* wood in the self-generated atmospheres (ACPR-S, ACPA-S), and under flowing air (ACPR-A) or N_2 (ACPA-N)

Parameter	ACPR-S	ACPR-A	ACPA-S	ACPA-N
S_{BET} (m^2/g)	2281	1638	968	1103
V_T (cm^3/g)	1.73	1.28	0.70	0.75
V_{micro} (%)	67	71	26	13
V_{meso} (%)	33	29	74	87
r (Å)	15	16	29	27

The results in Table 8 indicate that the BET surface area (S_{BET}) and the total pore volume (V_T) of the ACs developed from *Prosopis ruscifolia* were substantially higher than those for the *Parkinsonia aculeata*–based samples. For each wood species, the resulting ACs also evidence a reduction in the developed porous structures as stronger the oxidation conditions of the atmosphere, likely due to an enhanced development of porosity at the outer surface of the precursor's particle. It could arise from the more intensive carbon oxidation occasioned by the greater presence of oxygen involved in the thermal treatment stage. Besides, pronounced differences in the proportions of micro- and mesopores forming part of the ACs porous networks, depending on the wood used as precursor, are found. Whereas microporosity prevails in the ACs derived from *Prosopis ruscifolia*, contributions of mesopores to the total pore volume are predominant for the samples based on *Parkisonea aculeata*. The trend is reflected in the mean pore radius, with higher estimated values for the latter. Development of the porous structures is in line with yields shown in Figure 8. Lower yields are attained for the ACs derived from *Prosopis ruscifolia* possessing more developed porous structures.

3.7. Morphological Characteristics of the Wood Precursors and the Developed Activated Carbons

Typical SEM images of *Prosopis ruscifolia* wood and the activated carbons developed from this precursor under the self-generated atmosphere and flowing air are displayed in Figures 15 to 17 at different magnifications. The micrographs obtained for the activated carbons allow perceiving certain structural alterations, showing differences in morphological characteristics which depend on the activation atmosphere employed.

In Figures 15.a (magnification of 400x) and 15.b (magnification of 6000x), SEM images for the *P. ruscifolia* precursor are shown. From the figures, the compact structure of the precursor can be appreciated. The cellulosic structure corresponding to the fibre-tracheid and the multi-seriate woody ray are observed, but neither the vessels nor the cells corresponding to the longitudinal parenchyma can be noticed.

The morphological characteristics of the activated carbons developed from *P. ruscifolia* wood in the self-generated atmosphere (ACPR-S) can be observed in detail from the SEM images obtained at magnifications of 400x (Figure 16.a), 1600x (Figure 16.b), and 12000x (Figure 16.c). These samples present an organized structure which resembles a honeycomb. In Figure 16.a, different views of the material can be appreciated. The longitudinal view shows the cells of the radial parenchyma and some remains of the cells corresponding to the fibre-tracheid. On the other hand, from the cross section view, preservation of the vessels, surrounded by the residual structures of the cells which form the longitudinal parenchyma, is appreciated. The latter is better observed in Figure 16.b, which shows that the directional pattern of the vessels from the precursor remains almost intact and with no evidence of structural collapse. It is worth to mention that the vessels are authentic water and sap conducting tubes within the plant tissue and are formed by the junction of cells, namely the vascular element, which extends in the longitudinal direction of the tree. Chemical activation of the precursor in the self-generated atmosphere seems to preserve the original structure of the wood. Decomposition of lignin causes modifications of the original botanical structure and formation of volatile products, which render loss of the cohesive function of lignin (Khezami et al., 2005). In Figure 16.c, the residual structure of the cell wall may be recognized. It presents a considerable thickness with smooth walls, and well defined edges. Their internal surface is quite clean, with scarce deposits.

Figures 17.a to 17.c show the micrographs corresponding to the activated carbons developed from *Prosopis ruscifolia* under flowing air (ACPR-A) at different magnifications. The surface of the sample ACPR-A evidences modifications suffered by the material, likely as a consequence of the gasification produced due to the severe oxidation conditions employed in the synthesis (Figure 17.a).

In Figures 17.a and 17.b, it can be appreciated with different quality of detail that the original structure of the wood is not clearly recognizable, in opposition to previous observations depicted for the activated carbons developed in the self-generated atmosphere. In Figure 17.b, it can be seen that the original structure of the precursor has collapsed, likely as a result of the deformation of the vascular elements of the precursor. These present a flattened or elliptical shape and a variety of sizes, with undulating surfaces and a great quantity of carbonaceous deposits of different sizes. Also, the structure of the activated carbon presents remains of the structure of the tension fibres, which are typical in young trees.

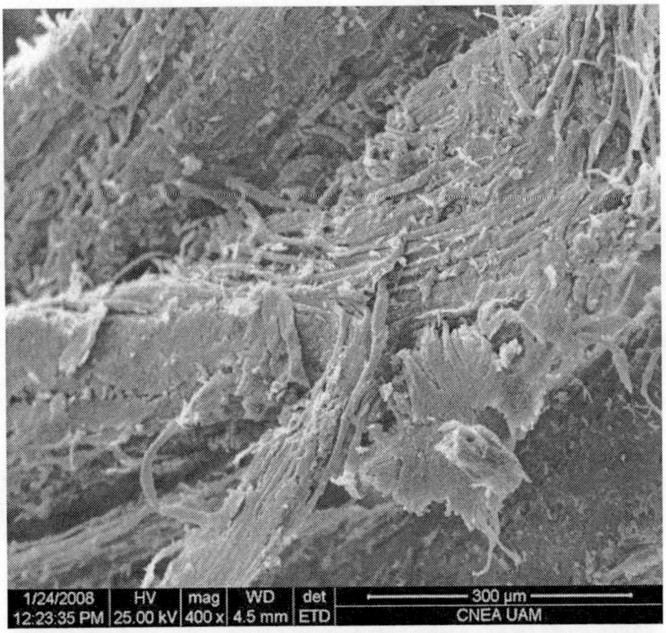

a)

b)

Figura 15. SEM micrographs of the cross section of *P. ruscifolia* employed as precursor for preparation of the activated carbons at magnifications of : a) 400x; b) 6000x.

a)

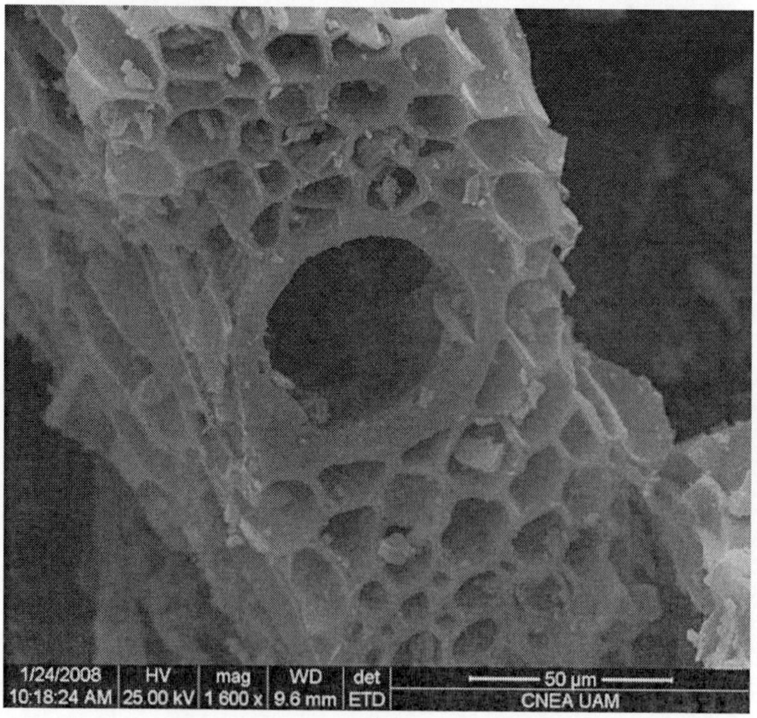

b)

Figure 16. (Continued).

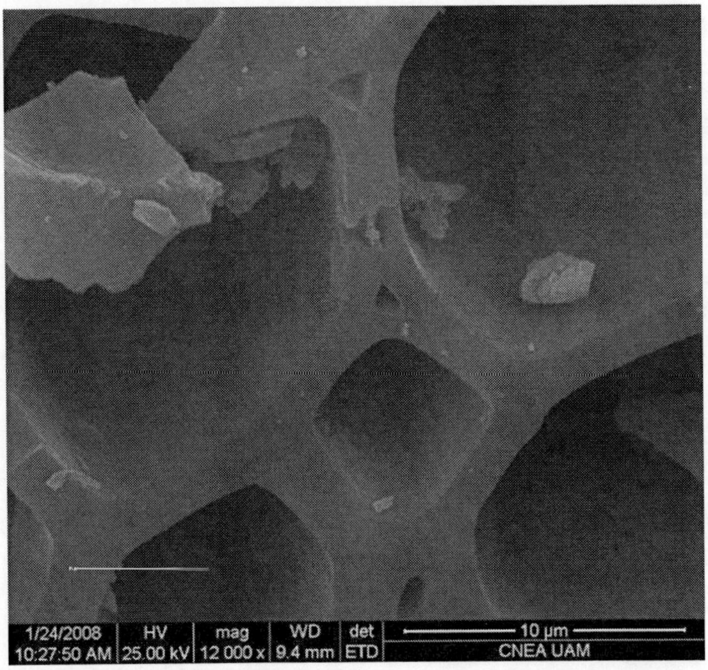

c)

Figura 16. a) SEM micrograph of the activated carbons developed from *Prosopis ruscifolia* under the self-generated atmosphere (400x). b) SEM micrograph of the activated carbons developed from *Prosopis ruscifolia* under the self-generated atmosphere (1600x).c) SEM micrograph of the activated carbons developed from *Prosopis ruscifolia* under the self-generated atmosphere (12000x).

In Figure 17.c the different layers of the cell wall may be recognized. Great deterioration in the walls surrounding the vessels may be observed, evidencing the reduction in their thickness in comparison to that displayed by the activated carbons developed in the self-generated atmosphere. This might be associated to the strongly oxidation atmosphere used in the preparation of ACPR-A, that could be responsible of the structural weakness of these activated carbons. Besides, the external surface presents cracks, also attributable to the more severe oxidative action caused by the flowing air during the thermal treatment stage of the activation process.

Similarly, illustrative SEM images of the surfaces of the *Parkinsonia aculeata* wood and of the activated carbons developed from this precursor under the self-generated atmosphere and flowing nitrogen are shown in Figures 18 to 20. Comparison of the cellular structure of the precursor (Figure 18) with those of the activated carbons (Figures 19 and 20) points to substantial changes occasioned by acid activation. Despite the changes, the honeycomb-shaped structure of the activated carbons (Figures 19.a-b) attests that the original morphology of the precursor is quite preserved. Although the surfaces of the ACs look fairly clean, disintegrated particles of different size and shape may be noticed. They could arise from thermal degradation of the lignocellulosic precursor, fragmentation of the cell walls, and/or from eventual condensation of carbonaceous material (Bonelli et al., 2001).

a)

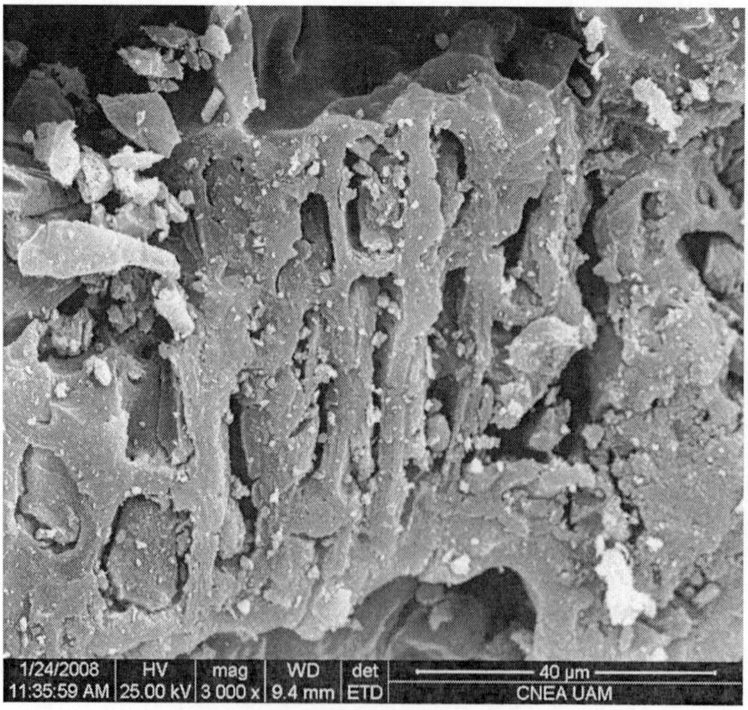

b)

Figure 17. (Continued).

c)

Figure 17. a) SEM image of the activated carbons developed from *Prosopis ruscifolia* under flowing air (400x). b) SEM image of the activated carbon developed from *Prosopis Ruscifolia* under flowing air (3000x). c) SEM image of the cross section of the activated carbon developed from *Prosopis Ruscifolia* under flowing air (12000x).

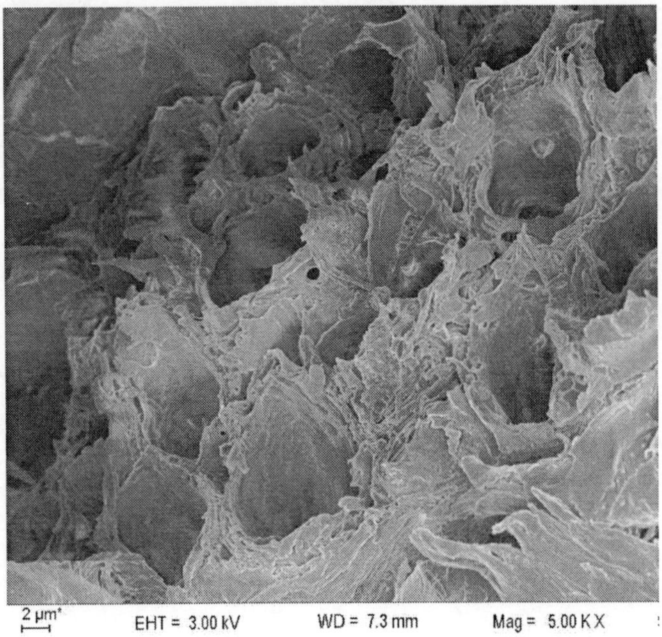

Figure 18. SEM micrograph of the cross section of *P. aculeata* employed as precursor for preparation of the activated carbons (5000x).

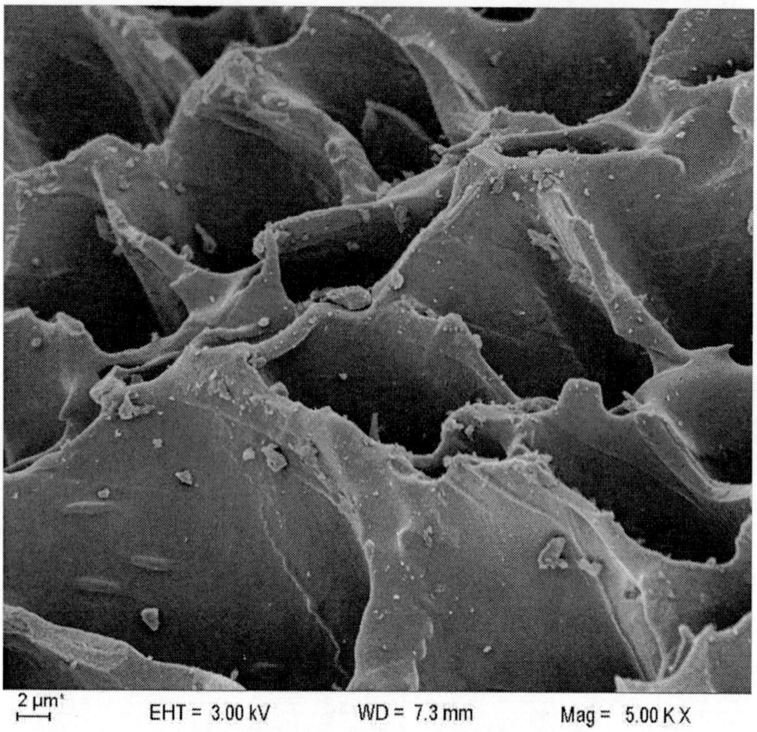

a)

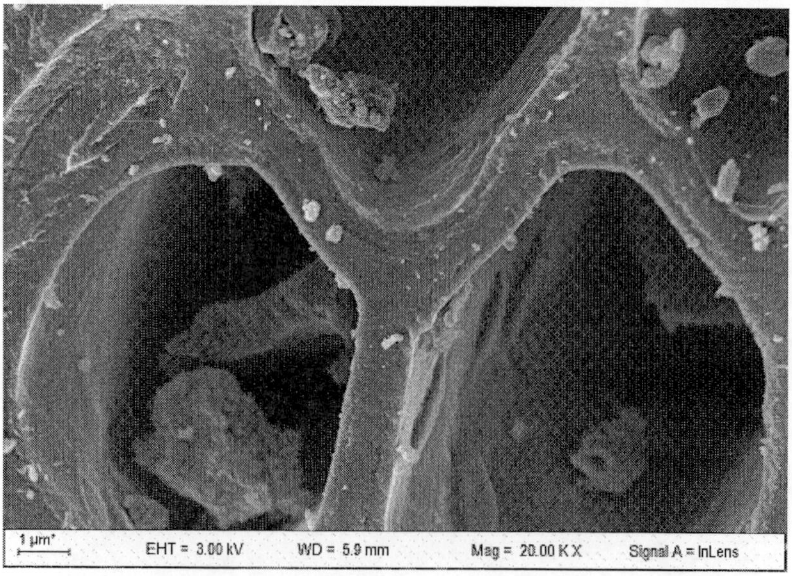

b)

Figura 19. a) SEM micrograph of the activated carbons developed from *Parkinsonia aculeata* under the self-generated atmosphere (5000x). b) SEM micrograph of the activated carbons developed from *Parkinsonia aculeata* under the self-generated atmosphere (20000x).

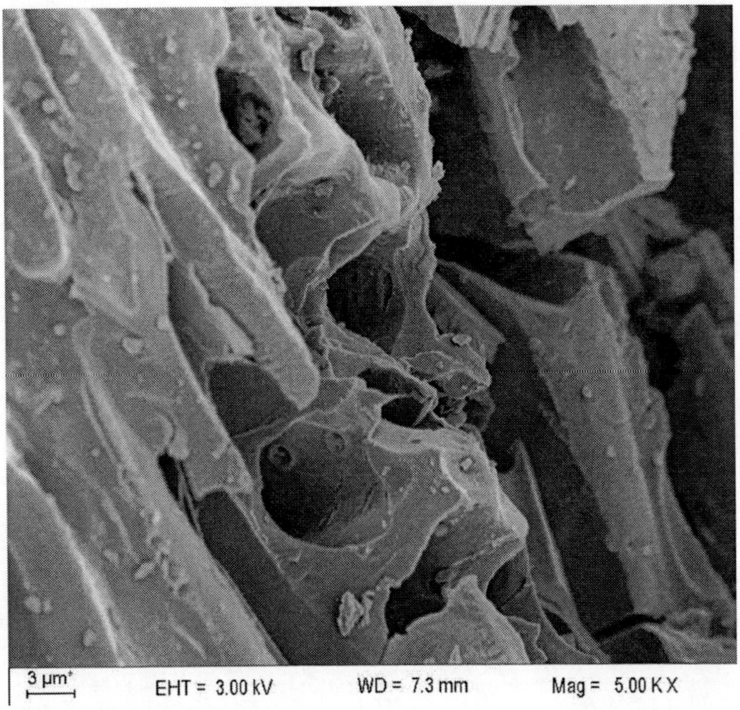

Figura 20. SEM micrograph of the activated carbons developed from *Parkinsonia aculeata* under flowing nitrogen (5000x).

3.8. Effectiveness of the Activated Carbons Developed from the Invasive Wood Species in Removal of Pollutants from Dilute Aqueous Solutions

In order to examine the potential use of the developed ACs for tertiary wastewater treatment, their ability in removing different pollutant species from dilute aqueous solutions was examined. For the ACs derived from *Prosopis ruscifolia*, their effectiveness in the adsorption of toxic metals and salicylic acid was tested. With respect to the latter, it is worthwhile to mention that pharmaceutical compounds have been recognized as a major group of emerging pollutants in recent years, due to their widespread occurrence in aquatic environments (Dubey et al., 2010). In turn, the adsorption capability of the *Parkinsonia aculeata* wood-based activated carbons toward nitrate ion, phenol, and methylene blue was investigated under pre-established equilibrium conditions. The effect of the sample's dose on removal of Cu[II] ions from solutions of 20 ppm for the ACs developed from *P. ruscifolia* wood sawdust is illustrated in Figure 21. As expected, increasing the dosage leads to enhance Cu[II] ion adsorption for both samples, regardless of the activation atmosphere used in the thermal treatment stage. Nevertheless, noticeable differences between ACPR-S and ACPR-A samples may be noticed for fixed doses. The ACPR-A developed under stronger oxidation conditions than the ACPR-S shows a better performance.

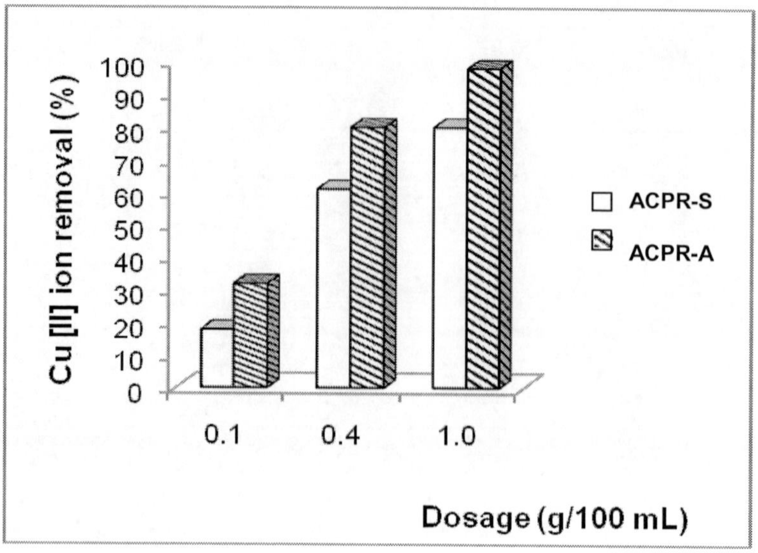

Figure 21. Cu[II] ion equilibrium adsorption at different doses of the activated carbons derived from *P. ruscifolia* wood under the self-generated atmosphere (ACPR-S) and flowing air (ACPR-A). Experimental conditions: $C_0 = 3.1x10^{-1}$ mM; pH = 6; t = 24 h; T = 28 °C.

Figure 22 shows the experimental Cu[II] ion adsorption isotherms determined for both ACs developed from *Prosopis ruscifolia* under the self-generated atmosphere and flowing air. Likewise, the adsorption isotherm of salicylic acid onto the ACs obtained from the same precursor under flowing air is illustrated in Figure 23.

In order to represent the experimental data and accounting for the shape of the isotherms, the Langmuir model was applied. It is described by the following equation (Cooney, 1999):

$$q_e = \frac{K_L X_{mL} C_e}{1 + K_L C_e} \tag{4}$$

Where X_{mL} is the maximum adsorption capacity, and K_L, the adsorption equilibrium constant. Model characteristic parameters were estimated by non-linear regression analysis for a 5% significance level, by minimising the following objective function:

$$O.F. = \Sigma \left(q_{ei,exp} - q_{ei,calc}\right)^2 \tag{5}$$

The appropriateness of the model to represent the experimental data was examined from the standard deviation (s), which was estimated according to:

$$s = 100 \left[\frac{\sum_{i=1}^{N} \left(q_{ei,exp} - q_{ei,est}\right)^2}{N - P} \right]^{1/2} \tag{6}$$

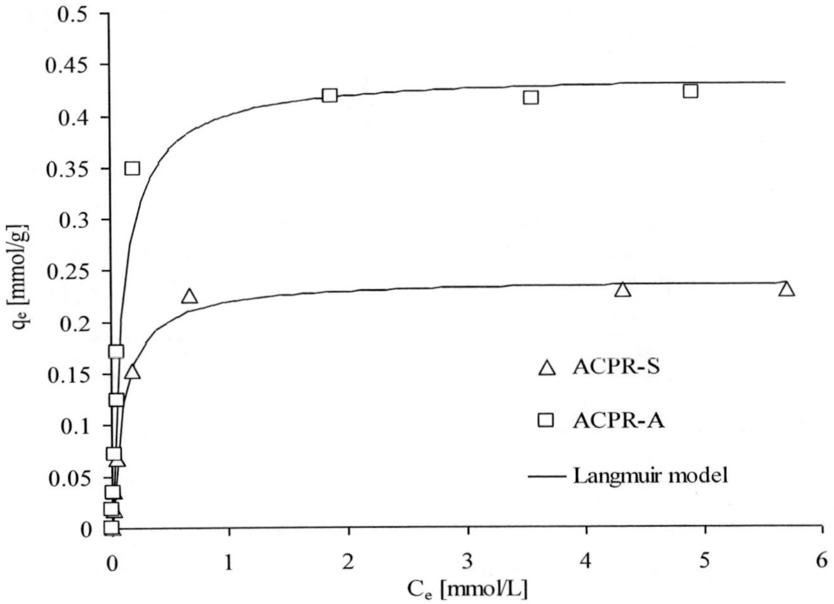

Figure 22. Adsorption isotherms of Cu[II] ion onto the activated carbons developed from *P. ruscifolia* wood under the self-generated atmosphere (ACPR-S) and flowing air (ACPR-A). Comparison between the experimental data (points) and predictions of the Langmuir model (solid lines). Experimental conditions: $C_0 = 8x10^{-2} – 6.3$ mM; pH = 6; T = 28 °C; carbon dose = 0.5 g/100 mL.

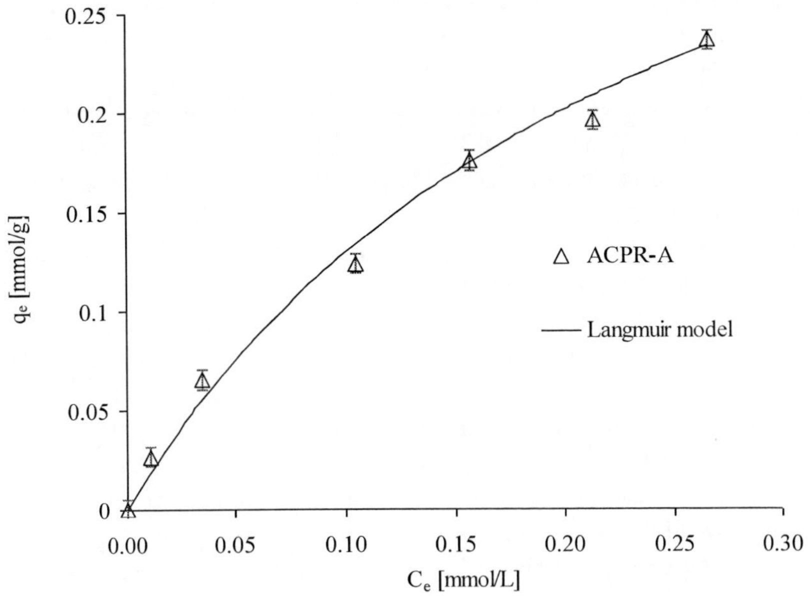

Figure 23. Adsorption isotherm of salicylic acid onto the activated carbons developed from *P. ruscifolia* wood under flowing air (ACPR-A). Comparison between the experimental data (points) and predictions of the Langmuir model (solid line). Experimental conditions: $C_0 = 7.2x10^{-2} – 1.45$ mM; pH = 4.5; T = 28 °C; carbon dose = 0.5 g/100 mL.

The estimated characteristic parameters of the Langmuir model along with the corresponding r^2 values are summarized in Table 9. Model predictions are shown in the same figures. As may be appreciated in Figures 22 and 23, the Langmuir model enables to represent properly the adsorption isotherms of the two solutes investigated onto the *Prosopis ruscifolia* – based ACs, with high r^2 values (Table 9).

Table 9. Langmuir model parameters estimated for the adsorption isotherms of Cu[II] ion or salicylic acid (AS) onto the activated carbons developed from *Prosopis ruscifolia* wood in the self-generated atmosphere (ACPR-S) and flowing air (ACPR-A)

System/ Parameter	X_m (mmol/g)	K_L (L/mmol)	s (%)	r^2
ACPR-S / Cu[II]	0.24	10.3	0.8	0.99
ACPR-A / Cu[II]	0.44	10.3	2.5	0.98
ACPR-A / AS	0.46	3.9	0.8	0.99

The results in Table 9 indicate that the maximum adsorption capacity of Cu[II] ion for the ACPR-A is near twice the one estimated for the ACPR-S. The found trend points to a predominant influence of the acidic/polar oxygen functional groups present on the surface of the ACs over that exerted by the porous structures. The ACPR-A with a relatively less developed porous matrix (Table 8) but with a higher amount of acidic/polar oxygen functionalities (Figure 11) shows a greater potentiality for Cu[II] ion removal, likely because dissociation or polarization of these functionalities renders to the surface of the ACs immersed in aqueous media anionic characteristics that favour their behaviour as cationic exchangers. The same trend has been reported for other activated carbons - metal ion systems (Basso et al., 2002; Basso and Cukierman, 2006). Based on present results and on predictions of a reported surface complexation model for copper adsorption onto activated carbons (Puziy et al., 2002), carboxylic groups and phenols could be the main functionalities involved in copper binding for the experimental conditions used in this study. Seemingly, carboxylic groups are involved in Cu[II] adsorption through the formation of a mono-dentate charged complex ($RCOOCu^+$), whereas phenolic groups appear to form a hydrated neutral complex (ROCuOH), the latter having a greater influence on adsorption from more concentrated solutions (Puziy et al., 2002). Values of the K_L parameter estimated for the ACs developed from *Prosopis ruscifolia* wood are similar, regardless of the activation atmosphere involved in the thermal treatment stage. On the other hand, maximum adsorption capacity of salicylic acid for the ACs developed under flowing air is similar to the value estimated for the metal ion, but noticeable differences in K_L parameter arise. The ACPA-A sample shows a reasonably acceptable performance, attaining a removal level higher than 80% for the highest initial concentration of the investigated range.

The ability of the activated carbons developed from *Parkinsonia aculeata* wood under the self-generated atmosphere and nitrogen flow in removing individually nitrate ion, phenol and methylene blue from dilute solutions of 1.6 mM initial concentration, at optimal pH value in each case, is comparatively illustrated in Figure 24. Pronounced differences in the adsorption between the organic compounds and nitrate ion may be noticed. While both carbons attain high removal percentages for the organics, their performance in removing nitrate is low. The adsorption effectiveness of the samples in decreasing order is methylene blue > phenol > nitrate ion. It is inversely related to the solubility of the solutes in water; the lower the solubility, the better the adsorption. Lower water solubility induces weaker solute-solvent bonds (Cooney, 1999). Thus, affinity of the solute for the carbon surface is favored and, consequently, a greater extent of adsorption is attained. Although the same trend holds for both activated carbons regardless of the activation atmosphere used in their preparation, a slightly better ability in removing the three species is found for those developed under flowing N_2, likely due to certain differences in the porous structure and/or surface chemistry between the samples.

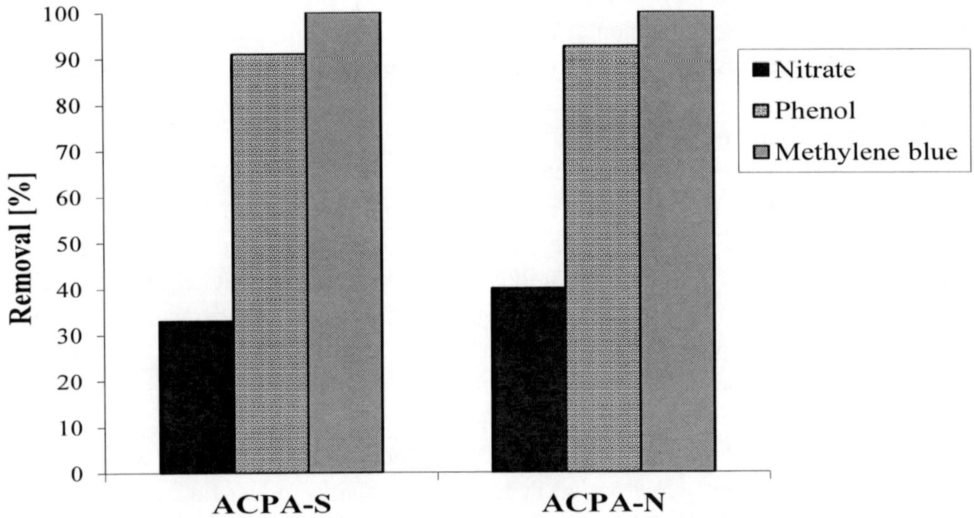

Figure 24. Effectiveness of the activated carbons developed from *Parkinsonia aculeata* wood under the self-generated atmosphere (ACPA-S) and flowing nitrogen (ACPA-N) in the removal of nitrate ion, phenol and methylene blue from dilute solutions (C_0 = 1.6 mM) at optimal pH values. Experimental conditions: T = 25 °C, dose = 1 g/100 mL.

CONCLUSION

Thermochemical processing of wood sawdust from two arboreal invasive species, *Prosopis ruscifolia* and *Parkinsonia aculeata,* has been examined in order to explore novel solutions to control their spread. The feasibility of converting wood sawdust from these species into products for energy applications and activated carbons has been investigated.

Yields of the products generated from rapid pyrolysis of the wood species, comprising bio-oil, non-condensable gas, and bio-char, have been determined from experiments conducted in a fixed-bed reactor at pre-established operating conditions. The wood species noticeably affects yields and distributions of the pyrolysis products. Pyrolysis of *Prosopis ruscifolia* generates a higher yield of bio-oil (~42%) and similar yields of the bio-char and gas (~ 28-29%), while similar yields of bio-oil and gas (~37-39%), higher than bio-char yield, arise from pyrolysis of *Parkinsonia aculeata* wood. Physico-chemical characteristics and higher heating value of the three kinds of products arising from pyrolysis of *Prosopis ruscifolia* indicate that they have potentialities for energy applications. The bio-oil, after being subjected to upgrading, could be employed as a liquid fuel with characteristics similar to those of fuel-oil, whereas the low heating value-gas evolved with pyrolysis course, mostly composed by CO_2, CO, CH_4 and H_2, might contribute to energy sustainability of the process. In turn, the bio-char (HHV≈ 27 MJ/kg) has potential as environmentally friendly solid bio-fuel.

Kinetic characterization for the pyrolysis of both wood species, from room temperature up to 500 ºC, reflects degradation dynamics of their main constituent biopolymers, namely cellulose, hemicellulose, and lignin. Application of an overall first-order decomposition model and the usual Arrhenius dependence of the reaction rate constant on temperature, in order to account for the complex reactions system taking place with pyrolysis course, enable to properly represent the experimental data over the whole temperature range investigated. The kinetic parameters evaluated point to differences in the pyrolytic reactivity between the wood species.

On the other hand, phosphoric acid activation of wood sawdust from the two invasive species at pre-established moderate conditions and *in situ* variation of the activation atmosphere involved in the thermal treatment stage of the process leads to high quality activated carbons, comparable to available commercial samples. Yields are similar or higher to those normally attained from conventional feedstocks. Higher yields are attained from *P.aculeata* (47%) than from *P. ruscifolia* (40%), regardless of the activation atmosphere.

Differences in main physico-chemical characteristics of the derived activated carbons, in turn governing their adsorptive behaviour, are found depending on the wood species used as a precursor and the activation atmosphere employed. The activated carbons developed from *Prosopis ruscifolia* wood in the self-generated atmosphere show higher BET surface area and total pore volume (2281 m^2/g, 1.7 cm^3/g) than the ones obtained under flowing air (1638 m^2/g and 1.3 cm^3/g), although the latter possess a higher total content of surface acidic/polar oxygen groups (2.2 meq/g) than the former (1.5 meq/g). Development of porous structures of the carbons derived from *P. ruscifolia* wood in both atmospheres is more pronounced than that attained for *P. aculeata*-based activated carbons, as revealed by the superior textural characteristics of the former, that agree with the lower yield attained. Besides, whereas microporosity prevails in the ACs derived from *P. ruscifolia,* the *P. aculeata*-based ACs show predominance of mesopores. The activated carbons developed from *P. aculeata* wood in the self-generated atmosphere possess BET surface area and total pore volume of 968 m^2/g, 0.70 cm^3/g, whereas values for those obtained under flowing N_2 are 1103 m^2/g and 0.75 cm^3/g. These samples have total contents of surface acidic/polar oxygen functionalities of 1.9 meq/g and 1.7 meq/g, respectively. Total contents of surface basic functionalities are very low or not detectable in all cases. Development of the surface chemistry of the resulting activated carbons apparently arises from the combination of the wood species and the activation atmosphere employed. *In situ* modification of the activation atmosphere affects both the total amount of acidic/polar surface oxygen groups and the relative proportions of the individual functionalities present on the surface of the derived activated carbons.

The ability of the activated carbons in removing representative toxic metal ions, oxo-anions, and organic compounds from dilute aqueous solutions mimicking water polluted with different contaminant species has been examined. As judged from determination of the adsorption isotherms of Cu[II] ion onto *P. ruscifolia*-based activated carbons and successful application of the Langmuir model, maximum sorption capacity (X_m) for the air-derived carbons (X_m= 0.44 mmol/g) almost duplicates the value for those obtained in the self-generated atmosphere (X_m= 0.24 mmol/g), pointing to a predominant effect of the surface acidic functionalities on metal sequestering behaviour. The air-derived carbons are also found acceptable for salicylic acid removal from dilute solutions. On the other hand, the *P. aculeata*-based activated carbons show high removal levels for phenol (~90%) and methylene blue (100%), but their ability in removing nitrate ion is low (≤ 40%). Removal effectiveness of these carbons is inversely related to water solubility of the solutes and is slightly affected by the activation atmosphere. Overall, effective activated carbons of potential applicability in tertiary wastewater treatment may be developed from the two invasive species, as suitable, alternative precursors, by phosphoric acid activation at moderate conditions.

ACKNOWLEDGMENTS

The authors gratefully acknowledged financial support from Consejo Nacional de Investigaciones Científicas y Técnicas (CONICET) and Universidad de Buenos Aires (UBA) from Argentina.

REFERENCES

Bandosz, T.J. 2006. *Activated carbon surfaces in environmental remediation.* Academic Press, London, UK.

Basso, M.C., Cerrella, E.G., Cukierman, A.L. 2002. Activated carbons developed from a rapidly renewable biosource for removal of cadmium (II) and nickel (II) ions from dilute aqueous solutions. *Ind. Eng. Chem. Res.* 41, 180-189.

Basso, M.C., Cukierman, A.L. 2005. *Arundo donax*-based activated carbons for aqueous-phase adsorption of volatile organic compounds. *Ind. Eng. Chem. Res.* 44, 2091-2100.

Basso, M.C., Cerrella, E.G., Buonomo, E., Bonelli, P., Cukierman, A.L. 2005. Thermochemical conversion of *Arundo donax* into useful solid products. *Energy Sources* 27, 1429-1438.

Basso, M.C., Cukierman., A.L. 2006. Wastewater treatment by chemically activated carbons from giant reed: Effect of the activation atmosphere on properties and adsorptive behavior. *Sep. Sci.Technol.* 41, 149-156.

Benaddi, H., Legras, D., Rouzaud, J.N., Beguin, F. 1998. Influence of the atmosphere in the chemical activation of wood by phosphoric acid. *Carbon* 36 (3), 306-309.

Blanco Castro, J., Bonelli, P.R., Cerrella, E.G., Cukierman, A.L. 2000. Phosphoric acid activation of agricultural residues and bagasse from sugar cane: influence of the experimental conditions on adsorption characteristics of activated carbons. *Ind. Eng. Chem. Res.* 39, 4166-4172.

Bonelli, P., Della Rocca, P., Cerrella, E.G., Cukierman, A.L. 2001. Effect of pyrolysis temperature on composition, surface properties and thermal degradation rates of Brazil nut shells. *Bioresource Technology* 76, 15-22.

Bonelli, P., Cerrella, E.G., Cukierman, A.L. 2003. Slow pyrolysis of nutshells: characterization of derived chars and of process kinetics. *Energy Sources* 25, 767-778

Bonelli, P., Buonomo, E., Cukierman, A.L. 2007. Pyrolysis of sugarcane bagasse and co-pyrolysis with an argentinean subbituminous coal. *Energy Sources Part A* 29, 731-740.

Bottani, E.J, Tascón, J.M.D. (Eds.) 2008. Overview of carbon materials in relation to adsorption, in: *Adsorption by Carbons*. Elsevier, Oxford, UK, 15-49.

Bozell, J.J. 2008. Feedstocks for the future - Biorefinery production of chemicals from renewable carbon. *Clean* 36, 41-47.

Bridgwater, A.V. 2011. Review of fast pyrolysis of biomass and product upgrading. *Biomass and Bionergy,* In press, doi:10.1016/j.biombioe.2011.01.048.

Butler, E., Devlin, G., Meier, D., McDonnell, K. 2011. A review of recent laboratory research and commercial developments in fast pyrolysis and upgrading. *Renewable and Sustainable Energy Reviews* 15, 4171-4186.

Channiwala, S.A., Parikh, P.P. 2002. A unified correlation for estimating HHV of solid, liquid and gaseous fuels. *Fuel* 81, 1051-1063.

Cochard, R., Jackes, B.R. 2005. Seed ecology of the invasive tropical tree *Parkinsonia aculeata. Plant Ecology* 180, 13-31.

Cooney, D.O. 1999. *Adsorption design for wastewater treatment,* CRC Press LLC, Boca Raton, USA.

Cordero, T., Marquez, F., Rodríguez Mirasol, T., Rodríguez, J.J. 2001. Predicting heating values of lignocellulosics and carbonaceous materials from proximate analysys. *Fuel* 80, 1567-1571.

Cukierman, A.L. 2007. Metal ion biosorption potential of lignocellulosic biomasses and marine algae for wastewater treatment. *Adsorption Science & Technology* 25 (3-4), 227-244.

De Celis, J., Amadeo, N.E., Cukierman, A.L. 2009. *In situ* modification of activated carbons developed from a native invasive wood on removal of trace toxic metals from wastewater. *Journal of Hazardous Materials* 161, 217-223.

Della Rocca, P.A., Cerrella, E.G., Bonelli, P.R., Cukierman, A.L. 1999. Pyrolysis of hardwood residues: On kinetics and chars characterization. *Biomass & Bioenergy* 16, 79-88.

Demirbas, A. 2009. Agricultural based activated carbons for the removal of dyes from aqueous solutions: A review. *Journal of Hazardous Materials* 167, 1-9.

Di Blasi, C., Signorelli, G., Di Russo, C., Rea, G. 1999. Product distribution from pyrolysis of wood and agricultural residues. *Ind. Eng. Chem. Res.* 38, 2216-2224.

Di Blasi, C. 2008. Modeling chemical and physical processes of wood and biomass pyrolysis. *Progress in Energy and Combustion Science* 34, 47-90.

Dubey, S.P., Dwivedi, A.D., Sillanpää, M., Gopal, K. 2010. *Artemisia vulgaris*-derived mesoporous honeycomb-shaped activated carbon for ibuprofen adsorption. *Chemical Engineering Journal* 165, 537-544.

Encinar, J.M., Beltrán, F.J., González, J.F., Moreno, M.J. 1997. Pyrolysis of maize, sunflower, grape and tobacco residues. *Journal of Chemical Technology and Biotechnology* 70 (4), 400-410.

Encinar, J.M., Gónzalez, J.F., Martínez, G., Romám, S. 2009. Jerusalem artichoke pyrolysis: Energetic evaluation. *Journal of Analytical and Applied Pyrolysis* 85, 294-300.

Fierro, V., Torné-Fernández, V., Celzard, A., Montané, D. 2007. Influence of the demineralisation on the chemical activation of Kraft lignin with orthophosphoric acid. *Journal of Hazardous Materials* 149, 126-133.

Garcia-Perez, M., Wang, X.S., Shen, J., Rhodes, M.J., Tian, F., Lee, W-J., Wu, H., Li, C. 2008. Fast pyrolysis of oil mallee woody biomass: Effect of temperature on the yield and quality of pyrolysis products. *Ind. Eng. Chem. Res.* 47, 1846-1854

Girgis, B.S., Ishak, M. 1999. Activated carbon from cotton stalks by impregnation with phosphoric acid. *Materials Letters* 39. 107-114.

Girgis, B.S., Soliman, A.M., Fathy, N.A. 2011. Development of micro-mesoporous carbons from several seed hulls under varying conditions of activation. *Microporous and Mesoporous Materials* 142, 518-525.

González, J.F., Ramiro, A., González-García, C.M., Gañán, J., Encinar, J.M., Sabio, E., Rubiales, J. 2005. Pyrolysis of almond shells. Energy applications of fractions. *Ind. Eng. Chem. Res.* 44, 3003-3012.

González, J., Kim, M.R., Buonomo, E., Bonelli, P., Cukierman, A.L. 2008. Pyrolysis of biomass from sustainable energy plantations: Effect of mineral matter reduction on kinetics and charcoal pore structure. *Energy Sources Part A* 30, 809-817.

Guo, Y., Rockstraw, D.A. 2007. Activated carbons prepared from rice hull by one-step phosphoric acid activation. *Microporous and Mesoporous Materials* 100, 12-19.

Gupta, R., Demirbas, A. 2010. *Gasoline, diesel, and ethanol biofuels from grasses and plants.* Cambridge University Press. New York, USA.

Ioannidou, O., Zabaniotou, A., Antonakou, E.V., Papasizi, K.M., Lappas, A.A., Athanassiou, C. 2009. Investigating the potential for energy, fuel, materials and chemicals production from corn residues (cobs and stalks) by non-catalytic and catalityc pyrolysis in two reactor configurations. *Renewable and Sustainable Energy Reviews* 13, 750-762.

Jagtoyen, M., Derbyshire, F. 1998. Activated carbons from yellow poplar and white oak by H_3PO_4 activation. *Carbon* 36, 1085-1097.

Keller, R.P., Geist, J., Jeschke, J.M., Kühn, I. 2011. Invasive species in Europe: Ecology, status, and policy. *Environmental Sciences Europe* 23:23 doi:10.1186/2190-4715-23-23.

Khezami, L., Chetouani, A.,Taouk, B., Capart, R. 2005. Production and characterisation of activated carbon from wood components in powder: Cellulose, lignin, xylan. *Powder Technology* 157, 48-56.

Mortensen, P.M., Grunwaldt, J-D., Jensen, P.A., Knudsen, K.G., Jensen, A.D. 2011. A review of catalytic upgrading of bio-oil to engine fuels. *Applied Catalysis A: General* 407, 1-19.

Nabais, J.V., Carrott, P., Ribeiro Carrott, M.M.L., Luz, V., Ortiz, A.L. 2008. Influence of preparation conditions in the textural and chemical properties of activated carbons from a novel biomass precursor: The coffee endocarp, *Bioresource Technology* 99, 7224-7231.

Nabais, J.V., Laginhasa, C., Carrott, P., Ribeiro Carrott, M.M.L. 2010. Thermal conversion of a novel biomass agricultural residue (vine shoots) into activated carbon using activation with CO_2. *Journal of Analytical and Applied Pyrolysis* 87, 8-13.

Özçimen, D., Ersoy-Meriçboyu, A. 2010. Characterization of biochar and bio-oil samples obtained from carbonization of various biomass materials. *Renewable Energy* 35 (6), 1319-1324.

Park, Y-K., Yoo, M.L., Lee, H.W., Park, S.H., Jung, S-Ch., Park, S-S., Kim, S-Ch. 2011. Effects of operation conditiones on pyrolysis characteristics of agricultural residues. *Renewable Energy.* In press, doi:10.1016/j.renene.2011.08.050

Patrick, J.W. (Ed.). 1995. *Porosity in carbons: Characterization and applications.* John Wiley & Sons, Inc. New York, USA.

Puziy, A.M., Poddubnaya, O.I., Martínez-Alonso, A., Suarez-García, F., Tascón, J.M.D. 2002. Synthetic carbons activated with phosphoric acid I, Surface chemistry and ion binding properties, *Carbon* 40, 1493-1505.

Puziy, A.M., Poddubnaya, O.I., Martínez-Alonso, A., Castro-Muñiz, A., Suarez-García, F., Tascón, J.M.D. 2007. Oxygen and phosphorus enriched carbons from lignocellulosic material. *Carbon* 45, 1941-1950.

Qu, T., Guo, W., Shen, L., Xiao, J., Zhao, K. 2011. Experimental study of biomass pyrolysis based on three major components: Hemicellulose, cellulose, and lignin. *Ind. Eng. Chem. Res.* 50, 10424-10433.

Ramos, M.E., Gonzalez, J.D., Bonelli, P.R., Cukierman, A.L. 2007. Effect of process conditions on physicochemical and electrical characteristics of denim-based activated carbon cloths. *Ind. Eng. Chem. Res.* 46, 1167-1173.

Ramos, M.E., Bonelli, P.R., Blacher, S., Ribeiro Carrott, M.M.L., Carrot, P.J.M., Cukierman, A.L. 2011. Effect of the activating agent on physico-chemical and electrical properties of activated carbon cloths developed from a novel cellulosic precursor. *Colloids and Surfaces A: Physicochemical and Engineering Aspects* 378, 87-93.

Richardson, D.M., Rejmánek, M. 2011. Trees and shrubs as invasive alien species - A global review. *Diversity and Distributions* 17, 788-809.

Rivera-Utrilla, J., Sánchez-Polo, M., Gómez-Serrano, V., Álvarez P.M., Alvim-Ferraz, M.C.M., Dias, J.M. 2011. Activated carbon modifications to enhance its water treatment applications. An overview. *Journal of Hazardous Materials* 187, 1-23.

Rodríguez-Reinoso, F. 2002. Production and application of activated carbon. In *Handbook of Porous Solids,* Wiley-VCH, Weinheim, Germany, 1766-1827.

Roskill Information Services. 2008. *The economics of activated carbon report,* 8th Edition, Roskill Information Services Ltd. http://www.roskill.com/reports/activated.

Saidur, R., Abdelaziz , E.A., Demirbas, A., Hossain, M.S., and Mekhilef ,S. 2011. A review on biomass as a fuel for boilers. *Renewable and Sustainable Energy Reviews* 15 (5), 2262-2289.

Schell, C., Riley, C., Peterson, G.R. 2008. Pathways for development of a biorenewables industry. *Bioresource Technology* 99 (12), 5160-5164.

Silvestre-Albero, A., Gonçalves, M., Itoh, T., Kaneko, K., Endo, M., Thommes, M., Rodríguez-Reinoso, F., Silvestre-Albero, J. 2012. Well-defined mesoporosity on lignocellulosic-derived activated carbons. *Carbon* 50, 66-72.

Telmo, C., Lousada, J. 2011. Heating values of wood pellets from different species. *Biomass and Bioenergy* 35 (7), 2634-2639.

U.S. Environmental Protection Agency. Wastewater Technology Fact Sheet. 2000. *Granular activated carbon: adsorption and regeneration.* U.S. EPA 832-F-00-017. Office of Water. Washington, D.C.

Vernersson, T., Bonelli, P., Cerrella, G., Cukierman, A.L. 2002. Arundo cane as a precursor for activated carbon preparation by phosphoric acid activation. *Bioresource Technology* 83, 95-104.

White, J.E., James Catallo, W., Legendre, B.L. 2011. Biomass pyrolysis kinetics: A comparative critical review with relevant agricultural residue case studies. *Journal of Analytical and Applied Pyrolysis* 91, 1-33.

Wongsiriamnuay, T., Tippayawong, N. 2010. Non-isothermal pyrolysis characteristics of giant sensitive plants using thermogravimetric analysis. *Bioresource Technology* 101, 5638-5644.

Yates, E.D., Levia, D.F., Williams, C.L. 2004. Recruitment of three non-native invasive plants into a fragmented forest in southern Illinois. *Forest Ecology and Management* 190, 119-130.

Yoder J., Galinato S., Granatstein D., García-Pérez, M. 2011. Economic tradeoff between biochar and bio-oil production via pyrolysis. *Biomass & Bioenergy* 35, 1851-1862.

Zhang, Q., Chang, J., Wang, T., Xu, Y. 2007. Review of biomass pyrolysis oil properties and upgrading research. *Energy Conversion and Management* 48, 87-92.

In: Invasive Species
Editors: Joaquín J. Blanco and Adrian T. Fernandes ISBN: 978-1-61942-761-7

Chapter 2

Invasive Rodents in the United States: Ecology, Impacts, and Management

Gary W. Witmer[1] and William C. Pitt[2]
[1]USDA/APHIS/WS National Wildlife Research Center, Fort Collins Colorado, US
[2]USDA/APHIS/WS National Wildlife Research Center, Hilo Hawaii, US

Abstract

Many invasive rodents have become established in the United States and its territories. The species include several species of *Rattus*, house mice (*Mus musculus*), Gambian giant pouched rats (*Cricetomys gambianus*), ground squirrels (*Spermophilus parryii*), nutria (*Myocastor coypus*) and marmots (*Marmota caligata*). While most were introduced accidently, some were introduced for food or fur. Additionally, some native species of rodents have been placed on islands, at least on a temporary basis, to study rodent species interactions. These rodents have caused serious impacts to native flora and fauna, agriculture, and other resources. They have caused the extinction or many species of birds on insular ecosystems. Although many methods are used to control or eradicate introduced rodents, rodenticides and traps are the main tools. Since the early 1990s, agencies have been eradicating rodents from various islands, primarily for conservation purposes. Of about 27 eradication attempts, 22 (81%) appear to have succeeded with only about 5 failures. For several islands, however, it is too early to determine if the attempted eradication has been successful or not. In the case of failed eradications, rapid re-invasion by rodents from nearby islands may be the reason. Numerous additional eradications are planned. We review the introduced rodent species, their impacts, and eradications, both successful and unsuccessful, that have occurred in the United States. Most eradications involved the use of the anticoagulant rodenticides diphacinone and brodifacoum. Rodenticides have been applied by hand-broadcast, bait station deployment, and aerial broadcast. We briefly review the strategies and methods used in eradication projects and the efforts to mitigate potential non-target and environmental impacts. Finally, we consider some of the remaining challenges in invasive rodent management and eradication in the United States, including the use of toxicants, land access, public attitudes, resource availability and monitoring difficulties.

INTRODUCTION

Many species of plants, microbes, and animals have been introduced around the world. Species are considered "alien" or "invasive" when they are not native to an area, but become established and cause, or are likely to cause, economic or environmental harm or harm to human health (NISC, 2008). Pimentel (2011) compiled information on the economic and environmental costs of introduced species around the world. Some vertebrate species were introduced purposefully, while others were introduced inadvertently or by escaping captivity. Purposeful introductions include animals used for food, fur, as work animals, or as companion animals. In some cases, they were introduced as a means of biological control (e.g., mongoose introduced to control rats). Game animals (including birds, mammals, and fish) have been widely introduced outside their native ranges to provide sporting opportunities and a source of game meat. Larger mammals, such as pigs and goats, were often introduced to islands by early explorers so that a supply of meat would be available to ships stopping on their long voyages. Those same ships were infested with rats and mice which, as a result, have colonized much of the world (Drake and Hunt, 2009).

Many species of terrestrial vertebrates have been introduced into parts of the United States and its territories (Witmer and Fuller, 2011). This includes all taxa of vertebrates:

- Mammals: 86 species
- Birds: 127 species
- Reptiles: 126 species
- Amphibians: 53 species
- Fish: 673 species

While many of these species were non-native to North America, many others were native, but had been moved from one region to another. An example is bullfrogs (*Rana catesbeiana*) native to eastern North America being moved to western North America where they have become an invasive species causing severe ecosystem disruption and even native species extinctions (Snow and Witmer, 2010).

The most common introductions are the commensal rodents, which have been widely introduced around the world (Long, 2003). However, it should be noted that many native rodent species occur worldwide. Approximately 40% of all mammalian species are rodents; this amounts to about 2,277 species (Wilson and Reeder, 2005). Native rodents have ecological, scientific, social, and economic values (Witmer et al., 1995). Rodents are important in seed and spore dispersal, pollination, seed predation, energy and nutrient cycling, the modification of plant succession and species composition, and as a food source for many predators. Additionally, some species provide food and fur for human uses. Rodents are also used extensively in medical research.

As invasive species, however, rodents are particularly problematic because they have many characteristics that make them very effective invaders. Rodent species have adapted to all life-styles: terrestrial, aquatic, arboreal, and fossorial (underground living). Most rodent species are small, secretive, nocturnal, adaptable, and have keen senses of touch, taste, and smell. In contrast to the normally small-sized body rodent, the capybara (*Hydrochaeris hydrochaeris*) of South America can reach 70 kg in mass. Rodents have excellent abilities to jump, climb, swim, and squeeze through small openings (Timm 1994a; Pitt et al., 2011a). For most species of rodents, the incisors continually grow throughout their lifespan, requiring constant gnawing to keep the incisors sharp and at an appropriate length. Additionally, rodents are known for their high reproductive potential. Many species have multiple litters per year with as many as 8-10 young per litter (Corrigan, 2001). Many species of rodents are omnivorous and can survive on a wide array of food types. Rodents can survive long periods of inclement weather or food shortages by storing foods for later use and by summer estivation and winter hibernation.

Numerous invasive rodents have become established in parts of the United States and its territories (Figure 1). The species include several species of *Rattus*, house mice (*Mus musculus*), Gambian giant pouched rats (*Cricetomys gambianus*), ground squirrels (*Spermophilus parryii*), nutria (*Myocastor coypus*), hoary marmots (*Marmota caligata*), and arctic ground squirrels (*Spermophilus parryii*). While most were introduced accidently, some were introduced for food or fur. Additionally, some native species of rodents (voles, *Microtus* spp. and deer mice, *Peromyscus* spp.) have been placed on islands, at least on a temporary basis, to study rodent species interactions (e.g., Crowell, 1983; Crowell and Pimm, 1976). Introduced rodents have caused serious impacts to native flora and fauna, agriculture, property, and other resources. Long (2003) reviewed the many rodent introductions around the world.

Figure 1. Introduced rodents, such as this roof rat, can cause extensive damage to island flora and fauna.

Several types of damage have been caused by rodent introductions to the United States (Hygnstrom et al., 1994; Witmer and Singleton, 2010). The substantial and world-wide loss of human food, both crops in field and stored foodstuffs, has been documented in several reviews (Meerburg et al., 2009a; Witmer and Singleton, 2010). In addition to consuming human foodstuffs, rodents also contaminate much more stored food through high levels of defecation and urination. Rodents also transmit many diseases to humans, companion animals, and livestock (Meerburg et al., 2009b). For example, the plague bacteria, *Yersinia pestis*---clausal agent of the Black Death which killed millions of humans worldwide in several pandemics---reached North America in the late 1800s via infected rats on ships arriving in California ports (Witmer, 2004).

Rodents can be prolific on islands where they have few or no predators. Their omnivorous foraging has led to the endangerment or extinction of numerous island species, especially bird species (Moors and Atkinson, 1984; Witmer et al., 1998; Veitch and Clout, 2002; Engeman et al., 2006; Towns et al., 2011). While their impacts to seabirds have been long known, invasive rodents also impact seeds and seedlings, invertebrates, sea turtle eggs and hatchings, and other resources (Witmer et al., 2007a; Caut et al., 2008; Angel et al., 2009; Towns et al., 2009; St. Clair, 2011; Drake et al., 2011). Most seabirds that nest on islands have not evolved to deal with mammalian predators and are very vulnerable to introduced rodents and other species introductions. In addition to direct effects, rodents can have many indirect effects on island resources through competition and trophic cascade effects (Russell, 2011). Invasive rodents have reached over 80% of the world's island groups where they have caused the demise of many endemic species (Atkinson, 1985). As a result, there has been a concerted worldwide effort to eradicate introduced rodents from islands with numerous successes (Howald et al., 2007; Witmer et al., 2011). These efforts have relied heavily on the use of rodenticides (Howald et al., 2007; Witmer et al., 2007b). In this chapter, we review the rodent introductions, impacts, and management strategies and methods used to reduce invasive rodent impacts and populations. We also review the invasive rodent eradication projects and methods used in the United States.

BIOLOGY AND ECOLOGY OF INTRODUCED RODENTS: SPECIES ACCOUNTS

Nutria or coypu *(Myocastor coypus)*, semi-aquatic rodents native to southern South America, are an invasive species having detrimental impacts mainly in the southern and eastern United States, but also in the Pacific Northwest. Nutria were introduced into the U.S. in 1899 for fur farming in California and later to several other states (Carter and Leonard, 2002). Nutria dispersals resulted primarily from releases by fur farmers, escapes during hurricanes or rising floodwaters, or as translocations in an attempt to control nuisance aquatic vegetation. Some states, such as Louisiana, continue to recognize nutria as a beneficial natural resource for fur and food, and manage populations for low densities – below presumed marsh vegetation damage thresholds. In other situations, such as at the Blackwater National Wildlife Refuge in Chesapeake Bay, Maryland, where nutria have caused excessive marsh damage, government agencies have implemented an eradication strategy (Kendrot and Sullivan, 2009). Nutria and the damage they cause to crops, canals, and wetlands have been well-described (Bounds et al., 2003; LeBlanc, 1994).

Generally, nutria have dark brown fur and weigh about 5-9 kg. At first glance at a nutria swimming, they can be mistaken for a beaver or a muskrat, both rodents native to North America. Female nutria are polyestrous and are sexually mature in approximately 5 months (LeBlanc, 1994). They are non-seasonal breeders capable of producing 3 to 4 litters a year with an average of 4 to 5 kits per litter. Nutria are voracious consumers of vegetation and are known to completely denude vegetation from areas where they feed before moving on. Their ease of mobility on land and in water makes them effective dispersers, posing significant challenges for resource managers.

Figure 2. Nutria damage to marsh vegetation in Maryland.

The ravenous appetite of these herbivores can cause damage to agricultural crops and aquatic vegetation, and can alter aquatic ecosystems (Figure 2). Crops damaged the most in the southeastern United States are rice and sugarcane, but other crops can be damaged as well: cereal grains, beets, peanuts, melons, and alfalfa (LeBlanc, 1994). In Louisiana, tens of thousands of acres of damaged marsh vegetation have been documented (Marx et al., 2004). Extensive marsh damage has also occurred in Maryland's Delmarva Peninsula (Kendrot and Sullivan, 2009). The areas damaged by nutria become permanent, open water ponds. Tidal and flooding impacts become more severe. The loss of marshland also removes habitat for native wildlife species such as waterfowl, wading birds, and muskrats (Bounds and Carowan, 2000; Southwick Associates, 2004). Finally, nutria burrowing habits can weaken irrigation structures and levees and they are a host for some diseases (LeBlanc, 1994).

Nutria populations and damage have been controlled mainly by private hunters and trappers. When nutria fur prices declined in the 1980s, damage in many areas became a great concern. In Louisiana, a method was devised to manage nutria damage and to supplement fur values with incentive payments to registered trappers and hunters of $4.00-5.00 per nutria tail. Unlike classic bounty systems, the program is intensively managed to target specific areas for population reduction; in 2003-2004, 332,596 nutria tails were collected in designated harvest areas by 346 participants (Marx et al., 2004). Traps, snares, shooting, and dogs have been used to remove nutria from the Delmarva Peninsula in Maryland (Kendrot and Sullivan, 2009). Rodenticides are rarely used for nutria control because of the potential hazards to non-target animals and water quality. Research continues to develop new methods to control nutria populations, such as multiple-capture live traps (Witmer et al., 2008) and improved attractants (Jojola et al., 2009).

Gambian giant pouched rats *(Cricetomys gambianus)* are native to a large area of central and southern Africa. They had become popular in the pet industry in some countries and became established on Grassy Key in the Florida Keys in 1999, following an escape or release by a pet breeder (Engeman et al., 2006; Perry et al., 2006). Despite a prolonged eradication effort, a free-ranging and breeding population remained on the island (Engeman et al., 2006; Engeman et al., 2007). There is a concern that if this rodent reaches the mainland, there could be damage to the Florida fruit industry because Gambian rats are known to damage numerous types of agricultural crops in Africa (Fiedler, 1994). Imported Gambian rats may also pose risks as reservoirs of monkey pox and other diseases. An outbreak of monkeypox occurred in the Midwestern United States in 2003 as a result of infected Gambian rats imported from Africa for the pet industry (Enserink, 2003). A climate-habitat modeling study suggested that their new range in North America could expand substantially were they to become established on the United States mainland (Peterson et al., 2006).

Figure 3. Gambian giant pouched rat captured in a cage trap on Grassy Key, Florida.

Gambian rats are gray brown in color and can reach a considerable size: about 2.8 kg in weight and about 1 m in length (Kingdon, 1974). Females produce 4 young per litter and can bear 8 or more litters per year (Ajayi, 1975). Because of their reproductive potential and their large size, they have been raised in captivity as a source of protein in Africa (Ajayi, 1975). Since free-ranging Gambian rats are new-comers to North America, relatively little is known about their biology, habitat use, impacts, and interactions with native species or about the most effective means to capture or control these rodents. Hence, current efforts are concentrating on use of traditional live trap capture methods (Figure 3) and rodenticides in bait stations (Engeman et al. 2007). Eradicating Gambian rats from Grassy Key has proven problematic because of the large number of private properties on the island, some of whose owners will not allow government employees on their property or will not allow the use of rodenticides on their property (Witmer and Hall 2011). It will be important to develop additional tools to manage or eradicate this species and other rodent invaders in the United States (Witmer et al., 2010; Witmer and Hall, 2011).

Norway rats *(Rattus norvegicus)* are native to a large part of Asia, but now occur worldwide with the exception of the polar regions (Long 2003). They were introduced to North America about 1775 in trans-Atlantic shipping (Brooks, 1973; Meehan, 1984). They are now completely established in both rural and urban areas throughout the United States, including Alaska and Hawaii. This species is one of the oldest and best known invasive vertebrates in the United States and is responsible for a variety of types of damage to crops and stored commodities (Jackson, 1977; Timm, 1994a). These rats spread rapidly and continuously across the country in shipped commodities, initially following wagon, riverboat, and rail routes. Areas with the least human traffic were the last to be reached.

The fur color of this rat is typically brown above and lighter brown gray below. The tail is sparsely haired and scaly and typically about the same length as the head and body. Norway rats generally weight about 500 g. One of the three common commensal rodent species on the North American continent, the Norway rat is closely tied to human settlements. Breeding may occur throughout the year. Females produce litters of 6-12 young and can bear 4-6 litters per year (Timm, 1994a). Gestation is about three weeks and animals reach sexual maturity in about three weeks (Timm, 1994a). Populations can expand rapidly when food, water, and habitat are available.

In farm settings, damage to stored food and grains, damage to garden crops, and predation on eggs and baby chickens is common. Grain consumption and fecal contamination is a common problem in commercial grain storage facilities (Jackson, 1977). Damage to roads, bridges, railroad track beds, and hydraulic structures may result from the burrowing activities and the associated soil loosening or flooding (Timm, 1994a). Structural damage in buildings results from gnawing and burrowing and may include damage to doors, window sills, and walls as well as to pipes and wiring. Insulation may be damaged or removed in the course of nest building. In urban areas, Norway rat populations are commonly associated with poor sanitation or accumulation of trash and food refuse in inner-city areas, although outdoor feeding of pets and wildlife often support suburban populations as well. Norway rats serve as reservoirs of a number of diseases that may affect humans and domestic animals, most commonly salmonellosis, leptospirosis, and trichinosis (Meehan, 1984). In areas with high rat populations in close association with humans, rat bites may occur, particularly to babies or young children.

Davis (1953) believed outdoor populations could be completely managed by environmental control and sanitation and demonstrated this repeatedly with experiments in Baltimore and New York City. However, Fall and Jackson (1998) contended that the political impossibility of maintaining diligence by urban residents and sustained support by public and private sectors has allowed Norway rat problems to continue unabated. Numerous products are available commercially to property owners for Norway rat control and extensive professional rodent control services are available through the structural pest control industry (Timm, 1994a; Corrigan, 2001).

Roof rats *(Rattus rattus)*, known also as black rats or ship rats, are native to a large portion of the Orient, probably throughout the Indo-Malayan region and through southern

China (Long, 2003). They also are now widespread worldwide, especially on tropical islands. In the United States, they occur along port and shore areas in southeastern and western North America and throughout Hawaii and tropical Atlantic and Pacific Ocean islands. Although known most commonly as a commensal species closely tied to man, this species, particularly in warmer areas, readily establishes in undeveloped areas, including native forests in Hawaii and on oceanic islands. According to Brooks (1973), roof rats were well established in Virginia in the early 1600s. They were well-established in North America's east coasts areas by the 1800s. They occur sporadically in warmer inland areas but rarely persist. However, a recent infestation discovered in urban Phoenix, Arizona raised concerns that the species could permanently establish in "islands" of suitable habitat and subsequently threaten crops and orchards (Nolte et al., 2003). In more temperate areas, they compete poorly with the larger and more aggressive Norway rat and occur mostly in port areas and generally indoors (Meehan, 1984).

The fur color is reddish-brown, brown or black, with the belly area being lighter or white. The tail is generally longer than the head and body. Adult roof rats weigh in the range of 150 to 250 g. As in Norway rats, breeding may occur throughout the year if resources are available and the pattern of breeding and the reproductive potential are similar between species. Roof rat females bear 3 or more litters with 5-8 young per litter each year (Marsh, 1994). Recently a variant of *Rattus rattus*, the Asian house rat, has been separated taxonomically as *Rattus tanezumi* (Musser and Carleton, 2005). Animals of both species are generally similar in appearance; however, *Rattus tanezumi* appears more variable and has a somewhat shorter tail. A chief distinguishing feature is a differing number of chromosomes between the two species, but this is of course not evident without use of special laboratory techniques and some authorities have not accepted the name change. *Rattus tanezumi* has recently been reported as a new invasive species in North America based on collections in California (James, 2006).

Like the Norway rat, the roof rat invades homes and structures, causing damage and contamination of stored food and commodities (Marsh, 1994). However, it readily adapts to field and forest habitats in tropical and semi-tropical areas causing damage to orchard, grain, and sugarcane crops. Because of their arboreal nature, roof rat can prey on adult birds, nestlings, and eggs under some circumstances and are recognized worldwide as the likely cause of rare bird extinctions in many island areas, including Hawaii (Munro, 1945; Atkinson, 1977; Pitt and Witmer, 2007). Black rats also eat snails and in Hawaii, they depredate the introduced predatory snail, *Euglandina rosea*, which has complicated management strategies to protect native tree snails (Meyer and Shiels, 2009). Roof rats also pose substantial threats to native plants through seed predation, as well as potentially aiding in the spread of non-native seeds via dispersal (Shiels, 2011; Shiels and Drake, 2011).

This species is also a reservoir for a number of diseases of humans and animals, but is most notorious for its role in bringing bubonic plague, the "Black Death," to 14th century Europe. The occurrence of bubonic plague in Hawaii during the period 1899 to 1958 was associated with this species (Tomich, 1986) as were the initial outbreaks in California in the early 1900s (Witmer, 2004).

Figure 4. Rat-proof bird nest box designed to protect endangered Hawaiian birds.

Control methods and materials are the same or similar to those used for Norway rat control. However, this species has been a particular target of recent efforts, both in the United States and in many other countries, to eradicate them from islands where seabirds or other desirable species are threatened by rat predation (Howald et al., 2007; Witmer et al., 2007a). Pitt et al. (2011b) recently developed and tested a nest box for endangered Hawaiian birds that prevents access by roof rats (Figure 4).

Polynesian rats (*Rattus exulans*), also known as Kiore or Pacific rats in Australasia, are a small tropical rat native to the Southeast Asia mainland that has spread throughout islands in the Pacific in conjunction with human settlement of the region (Matisoo-Smith and Robins, 2004). Although they do not occur on the United States mainland, they are well established on most tropical and semitropical islands (less than about 30° latitude) throughout the Pacific, including the Hawaiian Islands (Roberts, 1991). Polynesian rats are the smallest species (110-150 mm body length) in the genus *Rattus* and are slender (40-100 g) with relatively small feet and large ears. Like many rodents species, they are primarily nocturnal. Their fur is reddish brown on the dorsal surface and light gray on the belly. Polynesian rats may breed throughout the year and have up to 4 litters annually with 3-6 young in each (Jackson, 1965). They are sexually mature at 2 months and may have a life expectancy of around 1 year.

Polynesian rats have adapted to a wide range of habitats from forests to grasslands to agricultural crops, such as sugar cane. They are good climbers but do not swim so their dispersal to new islands is limited by human movement via ships and cargo (McCartney, 1970; Spenneman, 1997; Matisoo-Smith and Robins, 2004). They are opportunistic omnivores and their diets vary greatly by what is available by season and location so as to exploit locally abundant food sources (Kami, 1966; Kepler, 1967; Fall et al., 1971; Crook, 1973; Tobin and Sugihara, 1992; Sugihara, 1997; Rufaut and Gibbs, 2003). Predators of Polynesian rats include mongooses, cats, other larger rodents, and birds (Marshall, 1962). In addition, many Polynesian cultures consider rats to be a valuable food resource and rodents may have been introduced into new areas intentionally for food (Spenneman, 1997).

Polynesian rats are a significant agricultural pest throughout the Pacific region and they damage a variety of crops including rice, corn, macadamia nuts, sugarcane, coconut, cacao, pineapple, soybeans and root crops (Strecker, 1962; Tobin and Sugihara, 1992; Sugihara, 1997). Previous research documented the extensive effects of rat damage on sugarcane, but sugarcane production has largely been replaced by diversified agriculture in Hawaii (Pitt and Witmer, 2007). Rat damage has now shifted to high value seed crops (corn, soybean), and tropical fruits. Because Polynesian rats were spread through the Pacific Basin several thousand years ago, the impacts to the native flora and fauna are not readily apparent (Kepler, 1967; Crook, 1973; Rufaut and Gibbs, 2003; Meyer and Butaud, 2009). Polynesian rats are effective predators on sea birds, lizards, insects, and sensitive plant species that did not evolve with predation. Recent eradication efforts of Polynesian rats on islands have revealed these impacts as species recovery has occurred (Gibbs, 2009).

A variety of methods have been employed to reduce the effects of Polynesian rats on agriculture and the environment (Jackson, 1994). The primary successful methods have integrated rodenticides, alteration of cultural practices, and trapping (Sugihara, 1997). Rodenticides have been effectively used to reduce agricultural damage, protect forest birds, and protect seabird colonies. Previous attempts to control rat damage biologically have been unsuccessful and deleterious for other species. The most frequently cited failure is the introduction of mongoose to Hawaii in 1883 (Pitt and Witmer, 2007).

House mice *(Mus musculus* and *M. domesticus)* are native to southern Europe, northern Africa and Asia (Long, 2003). They now occur worldwide, including Antarctica, and are probably the most numerous and widespread mammalian species in the world next to humans (Witmer and Jojola, 2006). While house mice originated in the grasslands of Central Asia, they have followed humans around the world as stowaways on ships. They are very good invaders and probably reached to most parts of the world as stowaways on ships and cargo. House mice have remarkable abilities that have allowed them to be highly successful in many habitats around the world (Figure 5). Chief among these are their reproductive potential and their adaptability in different environments (Timm, 1994b; Witmer and Jojola, 2006).

House mice are small, slender rodents with fur that is grayish brown above and gray to buff underneath. This small (maximum mass of about 20 g for adults) and highly prolific animal is a continuous breeder in many situations; a female can produce 5-10 litters, each with 5-6 young, per year (Timm, 1994b). The young mature within about 3 weeks and soon become reproductively active. House mice are short-lived (generally less than one year) and have high population turn-over. In one study, 20 mice placed in an outdoor enclosure with abundant food, water, and cover, became a population of 2,000 in eight months (Corrigan, 2001).

Figure 5. House mice have amazing abilities which allow them to access almost any available areas or resources.

House mice cause many types of damage (Timm, 1994b; Witmer and Jojola, 2006). A major concern is the consumption and contamination of stored foods; it has been estimated that substantial amounts of stored foods are lost each year in this manner. Mice also consume and contaminate large amounts of livestock feed at animal production facilities. While mice generally live in close proximity to humans (Corrigan, 2001), sometimes feral populations occur. In these cases, the mice may damage many types of crops in the field, especially corn, cereal grains, and legumes. Australia has mouse "plagues" periodically resulting in enormous losses to stored crops and crops in the field (Brown et al., 2004). In buildings, a mouse infestation can be a considerable nuisance because of the noise, odors, and droppings. More importantly, they damage insulation and wiring (Hygnstrom, 1995). House fires have been caused by mice gnawing electrical wires; likewise, communication systems have been shut down for periods of time resulting in economic losses. Additionally, house mice are susceptible to a large number of disease agents and endo-parasites. Consequently, they serve as reservoirs and vectors of disease transmission to humans, pets, and livestock (Grantz, 1994). Important among these diseases are leptospirosis, plague, salmonella, lymphocytic choriomeningitis, and toxoplasmosis. Finally, when introduced to islands, mice can cause significant damage to natural resources, including both flora and fauna. For example, on Gough Island, mice fed on nestling albatross chicks (Cuthbert and Hilton, 2004). Additionally, Witmer (unpubl. data) documented seedling damage by house mice in a pen study. However, mice are subordinate to introduced rats so the impacts of mice not noticed when rats are also present on the island (Angel et al., 2009). This phenomenon was demonstrated by the large increase in mice abundance on Buck Island, USVI, after invasive roof rats were eradicated (Witmer et al., 2007a

A large number of methods and materials have been developed to help solve house mouse problems. In general, the use of multiple approaches and materials – integrated pest management – is more likely to reduce a mouse problem to a tolerable level (Witmer, 2007). The tools available and their proper use have been extensively reviewed (Brooks, 1973; Prakash, 1988; Timm, 1994b; Corrigan, 2001).

Hoary marmots (*Marmota caligata*) are native to parts of the mainland of North America, including portions of Alaska, USA, and the Yukon and Northwest Territories of Canada and southward through British Columbia, Canada, with fingers of their range extending along mountain ranges down into the states of Washington, Montana, and Idaho. The biology, ecology, and management of marmots were thoroughly reviewed by Armitage (2003). Of the several North American species of the *Marmota* genus, hoary marmots are the largest, weighing up to 7 kg. The fur is a grizzled-grey and more brownish towards the rear parts of the body and tail. The belly is whitish and the feet are all black. Hoary marmots live in loose colonies. These burrowing rodents are known for their sharp whistle warning of potential danger (such as a predator) given before scurrying to the safety of their burrow. The colony uses several burrow systems. The principle predators include badgers, wolverines, bears, wolves, and eagles.

The habitats used by hoary marmots are general high elevation, rocky areas and include alpine meadows, rocky talus slopes, and cliffs. These diurnal rodents are herbivores consuming a wide array of grasses and herbaceous vegetation; they will, however, consume some animal matter (Armitage, 2003). They put on significant body fat before entering winter hibernation. Often the entire colony will hibernate together in one burrow. Mating occurs soon after emergence from hibernation. There are 3-5 young per litter and only one litter per female per year (Armitage, 2003). Hence, marmots do not have the reproductive potential that rats and mice have.

Hoary marmots were introduced to Sud Island, Alaska, in about 1930 (Bailey, 1993), presumably to provide a source of food and/or fur for native and non-native people. While predatory, aquatic mammals (such as otters and seals) use the shoreline of the island, the marmots are the only terrestrial mammal occurring on the island. Additionally, this purposeful introduction has made marmots the only herbivore occurring on the island. The island is also home to about 23 species of birds, including seabirds, shorebirds, waterfowl, raptors, and passerines. Many of these bird species nest on the ground or in naturally-occurring burrows (e.g., rock crevasses).

The vegetation of Sud Island is comprised mainly of grasses, sedges, and several species of herbaceous plants (USDI, 2010). Hoary marmots consume large amounts of foliage, but also feed on flowers, seeds, berries, mosses, lichens, and roots (USDI, 2010). Marmots are known to impact plant communities in a number of ways, including species composition, species richness, and plant biomass (Del Moral, 1984; Semenov et al., 2001; Stallman and

Holmes, 2002; Armitage, 2003). On mainland areas, marmots are also known to damage trees (Anthony and Fisher, 1977; Swihart and Picone, 1994) and some of that damage may be related to scent-marking (Swihart, 1991a). Marmots can also cause significant damage to a variety of crops and hayfields (Swihart, 1991b; Bollengier, 1994). Their burrowing can also lead to increased soil erosion and can weaken dikes and building foundations (Bollengier, 1994). Marmots prefer southern to eastern slopes (Armitage, 2003), and on Sud Island, the most damage occurs on the eastern slopes (USDI, 2010). Marmots seemed less abundant on the steep western slopes, and hence, less damage is apparent (USDI, 2010). Marmots on Sud Island may also be competing with burrow-nesting birds for nest sites (USDI, 2010). Because marmots are known to consume some animal matter, it is possible that they are also doing some feeding on bird eggs and nestlings, but this has not been documented on Sud Island.

Marmot populations and damage can be managed in a number of ways, including exclusion, live-trapping and relocation, or killing with toxicants, shooting, or kill-traps (Dolbeer et al., 1991; Bollengier, 1994; Armitage, 2003). In general, research on repellents has not identified highly effective materials (Armitage, 2003) and laws and regulations affect which methods can be used in any given situation. An effort to eradicate the introduced hoary marmot population on Sud Island (see USDI, 2010) has been underway for the last year or two, but a number of factors have hindered success, including the steepness and rockiness of the slopes occupied by the marmots as well as the relatively low effectiveness of shooting and trapping which are the primary methods being used.

Arctic ground squirrels (*Spermophilus parryii*) are one of numerous species of ground squirrels that occur in North America. Arctic ground squirrels occur in Alaska, the Yukon and Northwest Territories of Canada and somewhat down into British Columbia, Canada. The biology, ecology, and management of ground squirrels were thoroughly reviewed by Yensen and Sherman (2003) and Banfield (1974). The arctic ground squirrel is the largest species of ground squirrel in North America with a weight of about 700 g. While much of the squirrel is tawny colored, the back is greyish to buffy-brown with white spots. Ground squirrels live in colonies and build extensive burrows, although permafrost often limits burrowing abilities. These ground squirrels are diurnal and their main predators include weasels, canids, and raptors.

Arctic ground squirrels occur on open tundra areas north of the treeline, but also within clearings in the northern forests. Arctic ground squirrels are mainly herbivores, but like most species of ground squirrels, will also each some animal matter, including insects and small vertebrates (Yensen and Sherman, 2003). They consume green foliage as well as seeds and berries. They also store large amounts of food to feed on after the long (often 7 months) hibernation period. There are 3-6 young per litter with only one litter per year (Yensen and Sherman, 2003) which is similar to the reproductive potential of marmots. Food resources seem to regulate population sizes, but additionally, winter survival can be low (Hubbs and Boonstra, 1997; Hubbs and Boonstra, 1998). Predation can also play a role in population dynamics (Hubbs and Boonstra, 1997; Karels et al., 2000).

While arctic ground squirrels occur widely across Alaska and on some islands near the mainland, they were introduced to a number of islands (Big Koniuji, Kavalga, Kodiak, Afognak, Unalaska, Amaknak Islands, and probably many others; Bailey, 1993) as early as 1895 as a source of food for introduced foxes and as a source of fur for people (Ebbert and Byrd, 2002). The foxes were introduced in the 1700s for the growing fur industry. Ironically, the squirrels were introduced as a new source of food for introduced foxes after the foxes had decimated the native seabird populations.

Arctic ground squirrels can have substantial effects on vegetation composition and biomass, especially in their high activity areas (Mallory and Heffernan, 1987; Frid and Turkington, 2001). Ground squirrel species occurring further to the south in North America, are known to cause substantial damage to crops, orchard trees, and rangeland forage (Askham, 1994). Their burrowing can weaken levees and undermine foundations and roadbeds (Askham, 1994). Additionally, some ground squirrels prey upon eggs and nestlings of ground-nesting birds (Askham, 1994). Hence, along with effects on island vegetation, arctic ground squirrels introduced to islands in Alaska could impact seabird populations (Ebbert and Byrd, 2002). The main methods of control of ground squirrel populations are through the use of traps, shooting, rodenticides, and burrow fumigants (Askham, 1994). Rodenticides have been proposed for use in control and/or eradication of arctic ground squirrels from some Alaskan islands, although care will be needed to reduce non-target animal losses (Ebbert and Byrd, 2002).

Other mainland rodent introductions have occurred as well. However, because many of these introductions were with rodent species native to North America, it is problematic at times to sort out which were actually introductions versus native species occurrences or natural range expansion. For example, the large aquatic rodent, North American beaver (*Castor canadensis*) was extirpated from much of North America as a result of trapping for the growing fur industry in the 1700s and 1800s (Baker and Hill, 2003). During the mid-1900s, they were re-introduced to many parts of their former range, but also to some new areas where they never occurred historically (Baker and Hill, 2003). Soon many of these populations were causing damage to commercial forest lands and other resources by tree girdling and cutting for food and to build dams (Miller and Yarrow, 1994; Baker and Hill, 2003). Extensive flooding often occurs after dams are built or when road culverts are blocked, damaging roadbeds and other resources. While beaver have a relatively low reproductive potential (one litter of 3-4 young per year), they are long-lived, reaching ages of 20 years in some cases (Miller and Yarrow, 1994). Populations are controlled by trapping, shooting, and by the use of pond levellers and culvert protective devices (Miller and Yarrow, 1994). Beaver were also introduced to some islands of the US and Canada (Long, 2003). Additionally, they have been introduced to Europe, Scandinavia, Russia, and Argentina (Long, 2003; Baker and Hill, 2003). Their range has greatly expanded in Chile and Argentina and extensive forest damage is occurring (Jaksic et al., 2002).

Fox and gray squirrels (*Sciurus niger* and *Sciurus carolensis*, respectively) are native to eastern North America, but have been introduced to parts of the far western United States and Canada (Edwards et al., 2003; Long, 2003). As per their eastern counterparts, introduced squirrels can damage trees, crops, wiring, and buildings as well as reducing tree seed recruitment (Jackson, 1994; Long, 2003; Krause et al., 2010). Once introduced, they are also known to out-complete some native populations of the tree squirrels (Long, 2003; Krause et al., 2010). Fortunately, tree squirrels have a relatively low reproductive potential (1 litter of 3 young per year; Jackson, 1994). Population control, as with beaver, is by use of traps and shooting (Jackson, 1994). Because fox squirrels introduced to California occur in urban/suburban settings, management options can be constrained by social considerations (Krause et al., 2010). Gray squirrels have also been introduced to Great Britain, South Africa, and Australia. Numbers of native red squirrels in Great Britain have declined substantially in numbers and range after the introduction and expansion of gray squirrel populations from North America (Long, 2003).

Muskrats (*Odontra zibethicus*) are one of the most widespread native species of rodent in North America, but populations have been introduced to some non-native areas such as parts of California (Long, 2003; Erb and Perry, 2003). Muskrats cause substantial amounts of crop damage, especially in areas where irrigated agriculture is practiced (Long, 2003). Damage also occurs from their burrowing activity and heavy foraging on aquatic vegetation (Miller, 1994; Erb and Perry, 2003; Long, 2003). Muskrats have a high reproductive potential (as many as 6 litters per year, each with 4-8 young; Miller, 1994). Additionally, muskrat populations are cyclic with high population numbers every 5-10 years (Miller, 1994). Control of populations is achieved by trapping and shooting and in some states, rodenticides can be used (Miller, 1994; Erb and Perry, 2003). Muskrats have been widely introduced across Europe and Asia as a source of fur (Long, 2003). In these regions, extensive damage to aquatic vegetation, and in some cases crops, has occurred (Long, 2003).

Finally, a few small populations of black-tailed prairie dogs (*Cynomys ludovicianus*), native to the prairies of the midwestern United States, were introduced to parts of Florida. Little has been reported on those populations or any damage that may be occurring. Additionally, a small population of capybara (*Hydrochaeris hydrochaeris*), native to South America, has become established in Florida. As with the prairie dog introductions, little is known about this population or any damage that may be occurring.

RODENT MANAGEMENT AND RESEARCH NEEDS

Many methods and tools have been developed and used to control rodent populations or to reduce the damage they cause (Table 1). Which methods are commonly used varies greatly from region to region around the world as well as between developed and undeveloped countries. Methods used also vary with regard to the type of management. With long-term management of rodent populations (such as in agricultural and urban/suburban settings) a greater variety of approaches are used, generally through an Integrated Pest Management (IPM) strategy (Witmer, 2007). So while traps and rodenticides are the mainstays of rodent population management, IPM also employs habitat management, exclusion and sanitation (Hygnstrom et al., 1994). On the other hand, if eradication of the invasive rodent species is the management goal, rodenticides are heavily relied upon, although traps may be used to some extent with the rodenticides. Some of the methods are highly regulated and regulations vary across political jurisdictions. The many methods used to manage rodent populations and damage have been described at length by Prakash (1988), Buckle and Smith (1994), Hygnsgtrom et al. (1994), and Caughley et al. (1998). In this chapter, we will only address traps and rodenticides in more detail.

A wide array of traps have been developed and used to manage rodents and many types are commercially available (Hygnstrom et al., 1994; Winn, 1986; Proulx, 1999). Trap types are subdivided into live traps and kill traps. With live traps the rodent becomes contained in a box or cage trap after tripping a treadle. Another type of live trap is the leg-hold trap which when tripped by the rodents paw springs the jaws of the trap to close tightly around the leg and hold the animal until the trapper arrives. Leg-hold traps are generally only used for larger rodent species such as nutria, muskrats, and beaver. Live traps often can be purchased at hardware or garden stores as well as through catalogues or websites. Leg-hold traps are generally obtained through websites or catalogues. Animals captured in live traps can be relocated (where regulations allow) to other locations or euthanized. An advantage of live traps is that non-target animals captured can often be released unharmed (Figure 6).

Table 1. Methods and techniques for rodent control that have been suggested, tested, or used to reduce rodent populations and damage around the world (from Witmer and Singleton, 2010)

Physical	Chemical	Biological	Other
Rodent proof construction	Baits/baiting systems	Fertility control	Appeasement
Passive barriers	Glues	Immunogens	Insurance
Electric barriers	Poison sprays	Habitat modification	Bounties
Drift fences	Poison moats	Cultural practices	Harvest
Trapping	Tracking powder	Crop timing	Compensation
Flooding burrows	Tracking greases, gel	Crop diversification, and species selection	
Drives	Repellents	Buffer crops	
Hunting	Attractants	Parasites	
Clubbing	Aversive agents	Diseases	
Frightening devices	Plant systematics	Predators	
Flame throwers	Sterilants	Ultrasonics	
Burrow destruction	Fumigation	Biosonics	
Habitat destruction	Psychotropic drugs	Resistant plants	
Harborage removal	Herbicides	Lethal genes	
Supplemental feeding	Poisons mixed with vehicle oil applied to flooded rice	Endophytic grasses	
Digging		Unpalatable plants	
Dogs together with flooding or digging			

Figure 6. Nutria in an experimental multiple capture live trap in Louisiana.

Kill traps cause the rapid death of the rodent by body constriction when the rodent trips the trap's trigger mechanism. The most common type of rodent kill trap is the snap trap. These are commonly sold at hardware and garden stores. Another type of kill trap is the Conibear trap used for larger rodent species. They can be purchased through websites or catalogues. Hygnstrom et al. (1994) provided good illustrations of various types of traps and directions for their proper and effective use. Effective trapping requires skill and practice. Using the proper type of trap for the situation, proper placement, and appropriate bait is very important to achieve a high level of trap success (i.e., a high capture rate). A disadvantage of kill traps is they can injure or kill non-target animals, including birds. Various types of traps are also used to monitor rodent populations. Rodent population monitoring is essential so that necessary management action can be taken before populations get very large at which point extensive damage to resources cannot be avoided.

Table 2. The main rodenticides used in the United States by category and percent active ingredient (from Witmer and Eisemann, 2007)

Acute Rodenticides
• Cholecalciferol (0.075%)
• Strychnine (0.5%)
• Zinc phosphide (2%)
• Bromethalin (0.01%)
Fumigants
• Aluminum phosphide (56%)
• Magnesium phosphide (56%)
• Acrolein (95%)
• Gas cartridges (variable)
1st Generation Anticoagulants
• Chlorophacinone (0.005%)
• Diphacinone (0.005%)
• Warfarin (0.025%)
• Pindone (0.025%)
2nd Generation Anticoagulants
• Bromadiolone (0.005%)
• Brodifacoum (0.005%)
• Difethialone (0.0025%)

Rodenticides are widely used in the United States as well other parts of the world. Because of their toxic nature and potential harm to people, pets, and livestock, rodenticides are carefully regulated by the United States Environmental Protection Agency (EPA) as well as by state agencies. There are many types of rodenticides and these vary by active ingredient as well as formulation (Table 2). These materials vary widely in their mode of action and in toxicity. The types and uses of rodenticides in the United States were reviewed by Witmer and Eisemann (2007). Their specific use for conservation purposes (i.e., the eradication of invasive rodents) was reviewed by Witmer et al. (2007b).

Proper training and careful use is required to safely use rodenticides so that they are effective in reducing rodent populations while minimizing the hazard to non-target animals. An EPA-approved product label provides considerable information on the product and its use, including: the registrant and EPA registration number(s), active ingredient and concentration, target species and settings in which it can be used, directions for use, storage and disposal requirements, precautionary statements, safety and environmental hazards, and threatened and endangered species considerations.

Both primary (direct consumption) and secondary hazards (consuming a poisoned rodent or poisoned non-target animal) can occur to non-target animals when rodenticides are used. Rodenticides such as brodifacoum (a second generation anticoagulant) are highly toxic, but also result in persistent residues in body tissues of animals that consume poisoned rodents (Witmer and Eisemann, 2007). There is a growing concern about the secondary effects of these residues in predatory animals (e.g., Thomas et al., 2011). The main safeguard for the safe use of rodenticides in the United States is carefully following the EPA label instructions for the product. Other considerations include the product used; when, where, and how it is applied; cleaning up spills promptly; and not using rodenticides where highly valued or protected wildlife occur (determined by scouting the area before use).

Additional research is needed to improve existing methods and to develop new methods for invasive rodent detection and control. More research is needed in both lethal and nonlethal means of resolving rodent damage situations (Witmer et al., 1995; Witmer and Singleton 2010). The research should include, but not be limited to, detection methods, new rodenticides, effective repellents, barrier development and improvement; biological control; fertility control; and habitat manipulation. Researchers also need to identify effective commercially-available rodenticide formulations for the various invasive rodents in each region of the country as Pitt et al. (2011c) has done for rats and mice in Hawaii. Another important research need is greater evaluation of the effectiveness of combinations of techniques, given that combinations could potentially be much more effective in the reduction of damage and may be more acceptable to the public.

INVASIVE RODENT ERADICATION

Since the early 1990s, federal and state agencies, along with conservation organizations, have been eradicating rodents from various islands in the United States, primarily for conservation purposes. Witmer et al. (2011) documented the attempted eradications of introduced rodents in the United States and its territories. Of about 27 eradication attempts, 22 (81%) appear to have succeeded with only about 5 failures. For several islands, however, it is too early to determine if the attempted eradication has been successful or not. Additionally, experimental rat eradication trials on 12 small islands in The Bay of Islands, Adak, Alaska, failed or rapid re-invasion occurred and those are not included in the list of more concerted eradication efforts as eradication methods were being investigated. In some cases, what appeared to be failed eradications may have resulted from rapid re-invasion by rats from nearby islands, suggesting the need to eradicate rats from groups of islands as an eradication unit. Genetic analyses of DNA from rats before and after eradications is helping sort out the issue of re-invasion versus failed eradication. Numerous additional eradications are underway or being planned. Most rodent eradications around the world have used the second-generation anticoagulant brodifacoum (Howald et al., 2007). In the United States, however, most eradications have used the first generation anticoagulant diphacinone. Initial rodent eradications used hand-broadcast and bait stations containing rodenticides, but in recent years, aerial broadcast via helicopter has become common. This allows rodent eradications on much larger and more rugged islands such as Rat Island, Alaska (2,700 ha; Witmer et al, 2011). Currently, the USDA Animal and Plant Health Inspection Service (APHIS) has two rodenticides registered with the EPA for island conservation purposes: one formulation of diphacinone pellets and two formulations of brodifacoum pellets (Witmer et al., 2007b; Figure 7).

A variety of mitigation measures are employed to reduce non-target hazards and environmental impacts. Examples include the rodenticide type, formulation, method and timing of baiting; placement of some non-target wildlife species in captivity until after the baiting operation; removal of rodent carcasses; and avoidance of bait placement in aquatic systems (Witmer et al., 2007b). In general, impacts to non-target species during invasive rodent eradications should be considered in terms of population-level effects, rather than the effects to individuals, and in terms of the "greater good' that is achieved from a successful eradication. While there will probably always be some losses of non-target animals, proper precautions should minimise such risk and allow for the rapid recovery of affected populations (Howald et al. 2005). Those involved with successful invasive rodent eradications on islands are often surprised at how rapidly the island's flora and fauna recover after rodents are removed (Witmer et al., 2007a).

PRECAUTIONARY STATEMENTS

HAZARDS TO HUMANS AND DOMESTIC ANIMALS

Keep away from humans, domestic animals and pets. If swallowed, this material may reduce the clotting ability of the blood and cause bleeding. Wear protective gloves when applying or loading bait. With detergent and hot water, wash all implements used for applying bait. Do not use these implements for mixing, holding, or transferring food or feed.

ENVIRONMENTAL HAZARDS

This pesticide is toxic to birds, mammals and aquatic organisms. Predatory and scavenging mammals and birds might be poisoned if they feed upon animals that have eaten bait.

PERSONAL PROTECTIVE EQUIPMENT (PPE)

Applicators and other handlers must wear:
-long sleeved shirt and long pants
-gloves
-shoes plus socks

For aerial application, in addition to the above PPE, loaders must wear protective eyewear or a face shield and a dust/mist filtering respirator (MSHA/NIOSH TC-21C).

USE RESTRICTIONS

It is a violation of Federal law to use this product in a manner inconsistent with its labeling. A copy of this label must be in the possession of the user at the time that the product is applied.

READ THIS LABEL: Read this entire label and follow all use directions and precautions.

IMPORTANT: Do not expose children, pets or other non-target animals to rodenticides. To help prevent accidents:
1) Keep children out of areas where this product is used or deny them access to bait by use of tamper resistant bait stations.
2) Store this product in locations out of reach of children, pets, and other nontarget animals.
3) Apply bait only according to the directions authorized.
4) Dispose of product container and unused, spoiled, or unconsumed bait as specified in the "**STORAGE AND DISPOSAL**" section.

(SEE RIGHT PANEL FOR ADDITIONAL USE RESTRICTIONS)

RESTRICTED USE PESTICIDE

DUE TO HAZARDS TO NON-TARGET SPECIES

For retail sale to and use only by Certified Applicators or persons under their direct supervision and only for those uses covered by the Certified Applicators certification.

For use by or in cooperation with government conservation agencies.

BRODIFACOUM-25D CONSERVATION

PELLETED RODENTICIDE BAIT FOR CONSERVATION PURPOSES

For control or eradication of invasive rodents in dry climates on islands or vessels for conservation purposes

ACTIVE INGREDIENT
Brodifacoum (CAS No. 56073-10-0) 0.0025%
INERT INGREDIENTS 99.9975%
TOTAL 100.0000%

KEEP OUT OF REACH OF CHILDREN

CAUTION

First Aid

If swallowed	-Call a physician or poison control center immediately for treatment advice. -Have person sip a glass of water if able to swallow. -Do not induce vomiting unless told to do so by a poison control center or doctor. -Do not give anything by mouth to an unconscious person.
If on skin or clothing	-Take off contaminated clothing. -Rinse skin immediately with plenty of water for 15-20 minutes. -Call a poison control center or doctor for treatment advice.
If inhaled	-Move person to fresh air. -If person is not breathing, call 911 or an ambulance, then give artificial respiration, preferably mouth-to-mouth if possible. -Call a poison control center or doctor for futher treatment advice.
If in eyes	-Hold eye open and rinse slowly and gently with water for 15-20 minutes. Remove contact lenses, if present, after the first 5 minutes, then continue rinsing eye. -Call a poison control center or doctor for treatment advice.

Have the product container or label with you when calling a poison control center or doctor, or when going for treatment.

For a medical emergency involving this product, call (877) 854-2494

NOTE TO PHYSICIAN: If swallowed, this material may reduce the clotting ability of blood and cause bleeding. If ingested, administer Vitamin K_1, intramuscularly or orally, as indicated in bishydroxycoumarin overdose. Repeat as necessary based on monitoring of prothrombin times.

USE RESTRICTIONS, (CONT)

This product may be used to control or eradicate Norway rats (*Rattus norvegicus*), roof rats (*Rattus rattus*), Polynesian rats (*Rattus exulans*), house mice (*Mus musculus*) or other types of invasive rodents on islands for conservation purposes, or on grounded vessels or vessels in peril of grounding.

This product may be applied using bait stations, burrow baiting, canopy baiting or by aerial and ground broadcast application techniques.

This product is to be used for the protection of State or Federally-listed Threatened or Endangered Species or other species determined to require special protection.

Do not apply this product to food or feed.

Treated areas must be posted with warning signs appropriate to the current rodent control project.

This product is for use in dry climates.

DIRECTIONS FOR USE

BAIT STATIONS: Tamper-resistant bait stations must be used when applying this product to grounded vessels or vessels in peril of grounding, or when used in areas of human habitation. Bait must be applied in locations out of reach of children, non-target wildlife, or domestic animals, or in tamper-resistant bait stations.
TO BAIT RATS: Apply 4 to 16 ounces (113 to 454 grams) of bait per placement. Space placements at intervals of 16 to 160 ft (about 5 to 50 meters). Placements should be made in a grid over the area for which rodent control is desired.
TO BAIT MICE: Apply 0.25 to 0.5 ounces (7 to 14 grams) of bait per placement. Space placements at intervals of 6 to 12 ft (about 2 to 4 meters). Larger placements, up to 2 ounces (57 grams) may be needed at points of very high mouse activity. Placements should be made in a grid over the area for which rodent control is desired.
FOR BOTH RAT AND MOUSE BAITING: Maintain an uninterrupted supply of fresh bait for at least 15 days or until signs of rodent activity cease. Where a continuous source of infestation is present, permanent bait stations may be established and bait replenished as needed.

Page 1 of 2
EPA Approved 03/18/08
EPA Reg. No. 56228-37

Figure 7. The EPA-approved label for a rodenticide designed for invasive rodent eradication on islands.

Planning and conducting a successful invasive rodent eradication from islands poses many challenges and should not be undertaken without a thorough commitment and adequate resources. The basic tenets of a successful eradication are: all individuals must be put at risk; animals must be removed faster than they can reproduce; and the risk of immigration must be zero (Parkes and Murphy, 2003). An eradication attempt that is 99% successful can ultimately result in 100% failure. Because of the large commitment of resources and public funds in eradication efforts, the potential for failure should be minimized. Planning and implementation components include:

- Preliminary monitoring and research,
- Feasibility of eradication,
- Regulatory compliance,
- Public information and communications media,
- Public support,
- Technical assistance and operations,
- Planning,
- Logistics,
- Procurement of equipment and other services,
- Monitoring and research,
- Staff recruitment and training,
- Implementation,
- Contingency planning,
- Follow up monitoring, and
- Implementation of a bio-security plan.

A number of challenges remain with invasive rodent management and eradication in the United States. Some of the challenges faced include the public and agency concerns about the use of toxicants and traps, land access (especially to private lands), public attitudes, resource availability, and detection and monitoring difficulties (Witmer and Hall, 2011; Witmer et al., 2011). Nonetheless, we will hopefully continue to relieve the burdens on insular and mainland ecosystems caused by rodent introductions. The flora and fauna of islands generally respond favorably and rapidly after invasive rodents are removed. Endemic, threatened, or endangered species can be, and have been, re-introduced after successful rodent eradications. For example, the endangered St. Croix ground lizard (*Ameiva polops*) was recently re-introduced to Buck Island in the U.S. Virgin Islands after the successful eradication of roof rats (*Rattus rattus*; Witmer et al. 2007a). The recent eradication of Polynesian rats and house mice from Cocos Island (a small island off of Guam) set the stage for the re-introduction of the endangered Guam rail, *Gallirallus owstonii* (Lujan et al., 2010).

CONCLUSION

Invasive rodents will continue to pose challenges to land and resource managers, commodity producers, and homeowners. Many tools are available to reduce rodent populations and associated damage. They should be used in a well thought out IPM approach.

Rodenticides will continue to be an important tool against rodents and their damage, but care must be exercised in their use. It is probably safe to assume that much of the public will continue to be leery of toxicant use. Hence, public education will be important to ensure continued availability of rodenticides. Continued technology development and transfer are essential to improve the effectiveness and safety of rodenticides and other methods used to control or eradicate invasive rodents.

Additionally, seabird populations, sea turtle populations and other island resources warrant protection from invasive rodents. The recovery of fauna and flora on uninhabited islands after a successful rodent eradication is particularly notable (Witmer et al., 2007a; Witmer et al., 2011). The significant impacts of introduced rodents on native flora and fauna have been repeatedly demonstrated. Invasive rodents are very adaptable, can exploit a wide array of resources as food and cover, and can increase reproduction very quickly when and where abundant resources exist (Macdonald et al., 1999). While invasive rodents will continue to pose challenges to land and resource managers, they can be controlled or even eradicated with a well-planned and adequately-supported effort using rodenticides and other tools. With proper planning, non-target losses will be minimal and these populations, along with other island resources, will often recover quickly after the invasive rodents have been removed.

REFERENCES

Ajayi, S. (1975). Observations on the biology, domestication, and reproductive performance of the African giant rat *Cricetomys gambianus* waterhouse in Nigeria. *Mammalia* 39, 343-364.

Angel, A., Wanless, R. M., & Cooper, J. (2009). Review of impacts of the introduced house mouse on islands in the Southern Ocean: are mice equivalent to rats? *Biological Invasions* 11, 1743-1754

Anthony, R. G. & Fisher, A. R. (1977). Wildlife damage in orchards – A need for better management. *Wildlife Society Bulletin* 5(3), 107-112.

Armitage, K. B. (2003). Marmots. In G. A. Feldhamer, B. C. Thompson, & J. A. Chapman (Eds.), *Wild Mammals of North America*. (Pp. 188-210), Baltimore and London: The Johns Hopkins University Press.

Askham, L. R. (1994). Franklin, Richardson, Columbian, Washington, and Townsend ground squirrels. In S. E. Hygnstrom, R. M. Timm, & G. E. Larson (Eds.), *Prevention and Control of Wildlife Damage*. (Pp. B-159 – B-163), Lincoln, Nebraska: University of Nebraska Cooperative Extension.

Atkinson, I. A. E. (1977). A reassessment of factors, particularly *Rattus rattus* L. that influenced the decline of endemic forest birds in the Hawaiian Islands. *Pacific Science* 31, 109-133.

Atkinson, I. (1985). The spread of commensal species of *Rattus*to oceanic islands and their effects on island avifaunas. In P. Moors (Ed.), *Conservation of Island Birds*. (Pp. 35-81), Cambridge, United Kingdom: International Council of Bird Preservation, Technical Bulletin No. 3.

Bailey, E. P. (1993). *Introduction of foxes to Alaskan islands – History, effects on avifauna, and eradication.* Washington D. C.: United States Department of the Interior Fish and Wildlife Service, Resource Publication 193. 53 pp.

Baker, B.W. & Hill, E.P. (2003). Beaver. In G. A. Feldhamer, B. C. Thompson, & J. A. Chapman (Eds.), *Wild Mammals of North America: Biology, Management, and Conservation.* (Pp. 288-310), Baltimore, Maryland: John Hopkins University Press.

Banfield, A. W. F. (1974). *The Mammals of Canada.* Toronto, Canada: University of Toronto Press. 433 pp.

Bollengier, Jr., R. M. (1994). Woodchucks. In S. E. Hygnstrom, R. M. Timm, & G. E. Larson (Eds.), *Prevention and Control of Wildlife Damage.* (Pp. B-183 – B-187), Lincoln, Nebraska: University of Nebraska Cooperative Extension.

Bounds, D. B. & Carowan, Jr., G. A. (2000). Nutria: a nonnative nemesis. *Transactions of the North American Wildlife and Natural Resources Conference* 65, 405-413.

Bounds, D. L., Sherfy, M. H., & Mollett T. A. (2003). Nutria. In G. A. Feldhamer, B. C. Thompson, & J. A. Chapman (Eds.), *Wild Mammals of North America: Biology, Management, and Conservation.* (Pp. 1119-1147), Baltimore, Maryland: John Hopkins University Press.

Brooks, J. E. (1973). A review of commensal rodents and their control. *CRC Critical Reviews in Environmental Control* 3(4), 405-453.

Brown, P. R., Davies, M., Singleton, G., & Croft, J. (2004). Can farm-management practices reduce the impact of house mouse populations on crops in an irrigated farming system? *Wildlife Research* 31, 597-604.

Buckle, A.P. & Smith, R.H. (1994). *Rodent Pests and Their Control.* Oxon, United Kingsom: CAB International. 405 pp.

Carter, J. & Leonard, B. P. (2002). A review of the literature on the worldwide distribution, spread of, and efforts to eradicate the coypu *(Myocastor coypus). Wildlife Society Bulletin* 30, 162-175.

Caughley, J., Bomford, M. Parker, B., Sinclair, R., Griffiths, J. & Kelley, D. (1998). *Managing Vertebrate Pests: Rodents.* Canberra, Australia: Bureau of Resource Sciences and Grains Research and Development Corporation. 130 pp.

Caut, S., Angulo, E., & Courchamp, F. (2008). Dietary shift of an invasive predator: rats, seabirds and sea turtles. *Journal of Applied Ecology* 45, 428-437.

Corrigan, R. M. (2001). *Rodent Control: A Practical Guide for Pest Management Professionals.* Richfield, Ohio: GIE, Inc. 355 pp.

Crook, I. G. (1973). The tuatara, *Sphenodon punctatus* Gray, on islands with and without populations of the Polynesian rat, *Rattus exulans* (Peale). *Proceedings of the New Zealand Ecological Society* 20, 115-120.

Crowell, K.L. (1983). Islands—insight or artifact?: population dynamics and habitat utilization in insular rodents. *Oikos* 27, 442-454.

Crowell, K.L. & Pimm, S.L. (1976). Competition and niche shifts of mice introduced onto small islands. *Oikos* 27, 251-258.

Cuthbert, R. & Hilton, G. (2004). Introduced house mice: a significant predator of threatened and endemic birds on Gough Island, South Atlantic Ocean? *Biological Conservation* 117, 483-489.

Davis, D. E. (1953). The characteristics of rat populations, *Quarterly Review of Biology* 28(4), 373-401.

Del Moral, R. (1984). The impact of the Olympic marmot on subalpine vegetation structure. *American Journal of Botany* 71(9), 1228-1236.

Dolbeer, R. A., Bernhardt, G. E., Seamans, T. W., & Woronecki, P. P. (1991). Efficacy of two gas cartridge formulations in killing woodchucks in burrows. *Wildlife Society Bulletin* 19, 200-204.

Drake, D. R., Bodey, T. W., Russell, J. C., Towns, D. R., Nogales, M., & Ruffino, L. (2011). Direct Impacts of Seabird Predators in Island Biota other than Seabirds. In C. P. H. Mulder, W. B. Anderson, D. R. Towns, & P. J. Bellingham (Eds.), *Seabird Islands: Ecology, Invasion, and Restoration.* (Pp. 91-132), New York, New York: Oxford University Press.

Drake, D. R. & Hunt, T. L. (2009). Invasive rodents on islands: integrating historical and contemporary ecology. *Biological Invasions* 11, 1483-1487.

Ebbert, S. E. & Byrd G. V. (2002). Eradications of invasive species to restore natural biological diversity on Alaska Maritime National Wildlife Refuge. In C. R. Veitch & M. N. Clout (Eds.), *Turning the tide: the eradication of invasive species.* (Pp. 102-109), Gland, Switzerland and Cambridge, UK: IUCN SSC Invasive Species Specialist Group.

Edwards, J., Ford M. & Guynn, D. (2003). Fox and gray squirrels. In G. A. Feldhamer, B. C. Thompson, & J. A. Chapman (Eds.), *Wild Mammals of North America.* (Pp. 248-267), Baltimore and London: The Johns Hopkins University Press.

Engeman, R. M, Witmer, G. W., Bourassa, J. B., Woolard, J. W., Constantin, B., Hall, P. T., Hardin, S., & Perry, N. D. (2007). The path to eradication of the Gambian giant pouched rat in Florida. In G. W. Witmer, W. C. Pitt, & K. A. Fagerstone (Eds.), *Managing Vertebrate Invasive Species: Proceedings of an International Symposium.* (Pp. 305-311), Fort Collins, CO: USDA/APHIS/Wildlife Services, National Wildlife Research Center.

Engeman, R., Woolard, J., Perry, N., Witmer, G., Hardin, S., Brashears, L., Smith, H., Muiznieks, B., & Constantin, B. (2006). Rapid assessment for a new invasive species threat: the case of the Gambian giant pouched rat in Florida. *Wildlife Research* 33,439-448.

Enserink, M. (2003). U.S. monkey pox outbreak traced to Wisconsin pet dealer. *Science* 300, 1639.

Erb, J. & Perry H. R. (2003). Muskrats. In G. A. Feldhamer, B. C. Thompson, & J. A. Chapman (Eds.), *Wild Mammals of North America.* (Pp. 311-348), Baltimore and London: The Johns Hopkins University Press.

Fall, M. W. & Jackson, W. B. (1998). A new era of vertebrate pest control? An introduction. *International Biodeterioration & Biodegradation* 42, 85-91.

Fall, M. W., Medina, A. B., & Jackson, W. B. (1971). Feeding patterns of *Rattus rattus* and *Rattus exulans* on Eniwetok Atoll, Marshall Islands. *Journal of Mammalogy* 52, 69-76.

Fiedler, L. A. (1994). Rodent Pest Management in Eastern Africa. *FAO Plant Production and Protection Paper* 123. Rome, Italy.

Frid, L. & Turkington, R. (2001). The influence of herbivores and neighboring plants on risk of browsing: a case study using arctic lupine (*Lupinus arcticus*) and Arctic ground squirrels (*Spermophilus parryii plesius*). *Canadian Journal of Zoology* 79, 874-880.

Gibbs, G. W. (2009). The end of an 80-million year experiment: a review of evidence describing the impact of introduced rodents on New Zealand's 'mammal-free' invertebrate fauna. *Biological Invasions* 11, 1587-1593.

Gratz, N. G. (1994). Rodents as carriers of disease. In A. P. Buckle & R. H. Smith (Eds.), *Rodent Pests and Their Control.* (Pp. 85-108), Wallingford, United Kingdom: CAB International.

Howald, G., Donlan, C. J., Galvan, J. P., Russell, J. C., Parkes, J., Samaniego, A., Wang, Y., Veitch, D., Genovesi, P., Pascal, M., Saunders, A., & Tershy, B. (2007). Invasive rodent eradication on islands. *Conservation Biology* 21(5), 1258-1268.

Howald, G., Faulkner, K., Tershy, B., Keitt, B., Gellerman, H., Creel, E., Grinnell, M., Ortega, S., & Croll, D. (2005). Eradication of black rats from Anacapa Island: biological and social considerations. In D. Garcelon & C. Schwemm (Eds.), *Proceedings of the 6th California Islands Symposium.* (Pp. 299-312), Arcata, California: National Park Service Technical Publication CHIS-05-01, Institute for Wildlife Studies.

Hubbs, A. H. & Boonstra, R. (1997). Population limitation in Arctic ground squirrels: effects of food and predation. *Journal of Animal Ecology* 66, 527-541.

Hubbs, A. H. & Boonstra, R. (1998). Effects of food and predators on the home-range sizes of Arctic ground squirrels (*Spermophilus parryii*). *Canadian Journal of Zoology* 76, 592-596.

Hygnstrom, S. E. (1995). House mouse damage to insulation. *International Biodeterioration and Biodegradation* 36, 143-150.

Hygnstrom, S., Timm, R., & Larson, G. (1994). *Prevention and control of wildlife damage.* Lincoln, Nebraska: University of Nebraska Cooperative Extension Service.

Jackson, J. J. (1994). Tree squirrels. In S. E. Hygnstrom, R. M. Timm, & G. E. Larson (Eds.), *Prevention and Control of Wildlife Damage.* (Pp. B-171 – B-175), Lincoln, Nebraska: University of Nebraska Cooperative Extension.

Jackson, W. B. (1977). Evaluation of rodent depredations to crops and stored products. *European Plant Protection Organization Bulletin* 7, 439-458.

Jaksic, F. M., Iriarte, J. A., Jiménez, J. E., & Martínez, D. R. (2002). Invaders without frontiers: cross-border invasions of exotic mammals. *Biological Invasions* 4, 157-173.

James, D. K. (2006). New rat species in North America. *Vector Ecology Newsletter* 37(1), 5.

Jojola, S., Witmer, G. W., & Burke, P. W. (2009). Evaluation of attractants to improve trapping success of nutria on Louisiana coastal marsh. *Journal of Wildlife Management* 73, 1414-1419.

Kami, H. T. (1966). Foods of rodents in the Hamakua District, Hawaii. *Pacific Science* 20, 367-373.

Karels, T. J., Byrom, A. E., Boonstra, R., & Krebs, C. J. (2000). The interactive effects of food and predators on reproduction and overwinter survival of arctic ground squirrels. *Journal of Animal Ecology* 69, 235-247.

Kendrot, S. & Sullivan, K. (2009). *An update of eradication progress for nutria in Chesapeake and Delaware Bay Wetlands.* Unpublished Report of USDA Wildlife Services, Cambridge, Maryland. 12 pp.

Kepler, C. B. (1967). Polynesian rat predation on nesting Laysan albatrosses and other Pacific seabirds. *Auk* 84, 426-430.

Kingdon, J. (1974). *East African Mammals. An Atlas of Evolution in Africa Volume 2, Part B (Hares and Rodents).* Chicago, Illinois: University of Chicago Press.

Krause, S.K., Kelt, D.A. & Van Vuren, D.H. (2010). Invasion, damage, and control options for Eastern fox squirrels. In R.M. Timm & K.A. Fagerstone (Eds.), *Proceedings of the 24nd Vertebrate Pest Conference.* (Pp.29-31), Davis, California: University of California.

LeBlanc, D. J. (1994). Nutria. In S. E. Hygnstrom, R. M. Timm, & G. E. Larson (Eds.), *Prevention and Control of Wildlife Damage*. (Pp. B-71 – B-80), Lincoln, Nebraska: University of Nebraska Cooperative Extension.

Long, J. (2003). *Introduced mammals of the world*. Collingwood, Australia: CSIRO Publishing. 590 pp.

Lujan, D.T., Vice, D.S., Guerrero, J.P. & Candaso, C.C. (2010). Rodent eradication on Cocos Island, Guam: integrating wildlife damage management, resort operations, and non-target concerns. In R.M. Timm & K.A. Fagerstone (Eds.), *Proceedings of the 24nd Vertebrate Pest Conference*. (Pp.9-13), Davis, California: University of California.

Macdonald, D., Mathews, F., Berdoy, M., & Deporter, M. (1999). The behavior and ecology of *Rattus norvegicus*: From opportunism to kamikaze tendencies. In G. Singleton, L. Hinds, H. Leirs, & Z. Zhang (Eds.), *Ecologically-based Management of Rodent Pests*. (Pp. 49-80), Canberra, Australia: Australian Centre for International Agricultural Research.

Mallory, F. F. & Heffernan, T. D. (1987). Floristic modification of low Arctic tundra by the Arctic ground squirrel, *Spermophilus parryii. The Canadian Field Naturalist* 101, 388-391.

Marsh, R.E. (1994a). Roof rats. In S. E. Hygnstrom, R. M. Timm, & G. E. Larson, (Eds.), *Prevention and Control of Wildlife Damage*. (Pp. B-125 – B-132), Lincoln, Nebraska: University of Nebraska Cooperative Extension.

Marshall, Jr., J. T. (1962). Predation and natural selection. In T. I. Storer (Ed.), *Pacific Island Rat Ecology: Report on a Study Made on Ponape and Adjacent Islands 1955-1958*. (Pp. 177-187), Honolulu, Hawaii: Bulletin 225, Bernice P. Bishop Museum.

Marx, J., Mouton, E., & Linscombe, G. (2004). *Nutria harvest distribution 2003-2004 and a survey of nutria herbivory damage in coastal Louisiana in 2004*. Baton Rouge, Louisiana: Fur and Refuge Division, Louisiana Department of Wildlife and Fisheries/Coastwide Nutria Control Program, CWPPRA Project (LA-03b). 45pp.

Matisoo-Smith, E. & Robins, J. H. (2004). Origins and dispersals of Pacific peoples: evidence from mtDNA phylogenies of the Pacific rat. *Proceedings of the National Academy of Science* 101, 9167-9172.

McCartney, W. C. (1970). Arboreal behavior of the Polynesian rat (*Rattus exulans*). *BioScience* 20, 1061-1062.

Meyer, J. Y. & Butaud, J. F. (2009). The impacts of rats on the endangered native flora of French Polynesia (Pacific Islands): drivers of plant extinction or coup de grâce species? *Biological Invasions* 11(7), 1569-1585.

Meyer, W.M. & Shiels, A.B. (2009). Black rat (*Rattus rattus*) predation on non-indigenous snails in Hawaii: complex management implications. *Pacific Science* 63, 339-347.

Meehan, A. P. (1984). *Rats and Mice*. East Grinstead, UK: Rentokil Ltd. 383 pp.

Meerburg, B. G., Singleton, G. R., & Leirs, H. (2009a). The Year of the Rat ends – time to fight hunger. *Pest Management Science* 65, 351-352.

Meerburg, B., Singleton, G., & Kijlstra, A. (2009b). Rodent-borne diseases and their risks for public health. *Critical Reviews in Microbiology* 35, 221-270.

Miller, J. E. (1994). Muskrats. In S. E. Hygnstrom, R. M. Timm, & G. E. Larson (Eds.), *Prevention and Control of Wildlife Damage*. (Pp. B-61 – B-69), Lincoln, Nebraska: University of Nebraska Cooperative Extension.

Miller, J. E. & Yarrow, G. K. (1994). Beavers. In S. E. Hygnstrom, R. M. Timm, & G. E. Larson (Eds.), *Prevention and Control of Wildlife Damage*. (Pp. B-1 – B-11), Lincoln, Nebraska: University of Nebraska Cooperative Extension.

Moors, P. J. & Atkinson, I. A. E. (1984). Predation on seabirds by introduced animals, and factors affecting its severity. In P. J. Moors (Ed.), *Conservation of Island birds: case studies for the management of threatened island birds*. (Pp. 667-690), Cambridge, United Kingdom: International Council for Bird Preservation Technical Publication No. 2.

Munro, G. C. (1945). Tragedy in bird life. *Elapaio* 5, 48-51.

Murphy, J. & Ohashi, T. (1993). *Report of rat eradication operations conducted under specific emergency exemption to use weatherblok containing brodifacoum and assault containing bromethalin in a field situation on Rose Atoll National Wildlife Refuge, American Samoa*. Unpublished Report of USDI Fish and Wildlife Service, Honolulu, Hawaii. 18 pp.

Musser, G. G. & Carleton, M. D. (2005). Superfamily Muroidea. In D. E. Wilson & D. M. Reeder (Eds.), *Mammal Species of the World: A Taxonomic and Geographic Reference (3rd Ed)*. (Pp. 894-1531), Baltimore and London: Johns Hopkins University Press.

National Invasive Species Council. (2008). *2008-2012 National Invasive Species Management Plan*. Washinton, D.C.: National Invasive Species Council. 35 pp.

Nolte, D. L., Bergman, D., & Townsend, J. (2003). Roof rat invasion of an urban desert island. In G. R. Singleton, L. A. Hinds, C. J. Krebs, & D. M. Spratt (Eds.), *Rats, Mice, and People: Rodent Biology and Management*. (Pp. 481-484), Canberra, Australia: Australian Centre for International Agriculture Research.

Parkes, J. & Murphy, E. (2003). Management of introduced mammals in New Zealand. *New Zealand Journal of Zoology* 30, 335-359.

Perry, N. D., Hanson, B., Hobgood, W., Lopez, R. L., Okraska, C. R., Karem, K., Damon, I. K., & Carroll, D. S. (2006). New invasive species in southern Florida: Gambian rat. *Journal of Mammalogy* 87, 262-264.

Peterson, A. T., Papeş, M., Reynolds, M. G., Perry, N. D., Hanson, B., Regnery, R. L., Hutson, C. L., Muizniek, B., Damon, I. K., & Carroll D. S. (2006). Native-range ecology and invasive potential of *Cricetomys* in North America. *Journal of Mammalogy* 87, 427-432.

Pimentel, D. (2011). *Biological Invasions: Economic and Environmental Cost of Alien Plant, Animal, and Microbe Species*. Boca Raton, Florida: CRC Press. 449 pp.

Pitt, W. C., Sugihara, R. T., Driscoll, L. C., & Vice, D. S. (2011a). Physical and behavioural abilities of commensal rodents related to the design of selective rodenticide bait stations. *International Journal of Pest Management* 57(3), 189-193.

Pitt, W. C., Driscoll, L. C., & VanderWerf, E. A. (2011b). A rat-resistant artificial nest box for cavity-nesting birds. *Human-Wildlife Interactions* 5(1), 100-105.

Pitt, W. C., Driscoll, L. C., & Sugihara, R. T. (2011c). Efficacy of Rodenticide Baits for the Control of Three Invasive Rodent Species in Hawaii. *Archives of Environmental Contamination and Toxicology* 60, 533-542.

Pitt, W. C. & Witmer, G. W. (2007). Invasive Predators: a synthesis of the past, present, and future. In A. M. T. Elewa (Ed.), *Predation in Organisms -- a Distinct Phenomenon*. (Pp. 265-293), Heidelberg, Germany: Springer Verlag.

Prakash, I. (1988). *Rodent Pest Management*. Boca Raton, Florida: CRC Press. 496 pp.

Proulx, G. (1999). *Mammal Trapping*. Sherwood Park, Alberta, Canada: Alphs Wildlife Reseaerch & Management, Ltd.. 231 pp.

Roberts, M. (1991). Origin, dispersal routes, and geographic distribution of *Rattus exulans*, with special reference to New Zealand. *Pacific Science* 45, 123-130.

Rufaut, C. G. & Gibbs, G. W. (2003). Response of a tree weta population (*Hemideina crassidens*) after eradication of the Polynesian rat from a New Zealand island. *Restoration Ecology* 11, 13-19.

Russell, J. C. (2011). Indirect Effects of Introduced Predators on Seabird Islands. In C. P. H. Mulder, W. B. Anderson, D. R. Towns, & P. J. Bellingham (Eds.), *Seabird Islands: Ecology, Invasion, and Restoration*. (Pp. 261-279), New York, New York: Oxford University Press.

Semenov, Y.S., Rammousse, R. & Le Barre, M. (2001). Impact of black-capped marmot on floristic diversity of arctic tundra in Northern Siberia. *Arctic, Antarctic, and Alpine Research* 33, 204-210.

Shiels, A.B. (2011). Frugivory by introduced black rats (*Rattus rattus*) promotes dispersal of invasive plant seeds. *Biological Invasions* 13, 781-792.

Shiels, A.B. & Drake, D.R. (2011). Are introduced rats (*Rattus rattus*) both seed predators and dispersers in Hawaii? *Biological Invasions* 13, 883-894.

Snow, N.P. & Witmer, G.W. (2010). American bullfrogs as invasive species: a review of the introduction, subsequent problems, management options, and future directions. In R.M. Timm & K.A. Fagerstone (Eds.), *Proceedings of the 24nd Vertebrate Pest Conference*. (Pp.86-89), Davis, California: University of California.

Southwick Associates. (2004). *Potential economic losses associated with uncontrolled nutria populations in Maryland's portion of the Chesapeake Bay*. Ferandina Beach, Florida: Southwick Associates. 17 pp.

Spenneman, D. H. R. (1997). Distribution of rat species (*Rattus* spp.) on the atolls of the Marshall Islands: past and present dispersal. *Atoll Research Bulletin* 446, 1-18.

St Clair, J. J. H. (2011). The impacts of invasive rodents on island invertebrates. *Biological Conservation* 144, 68-81.

Stallman, E. L. & Holmes, W. G. (2002). Selective foraging and food distribution of high-elevation yellow-bellied marmots (*Marmota flaviventris*). *Journal of Mammology* 83(2), 576-584.

Strecker, R. L. (1962). Economic relations. In T. I. Storer (Ed.), *Pacific Island Rat Ecology: Report on a Study Made on Ponape and Adjacent Islands 1955-1958, Bulletin 225*. (Pp. 200-218), Honolulu, Hawaii: Bernice P. Bishop Museum.

Sugihara, R. T. (1997). Abundance and diets of rats in two native Hawaiian forests. *Pacific Science* 51, 189-98.

Swihart, R. K. (1991a). Modifying scent-marking behavior to reduce woodchuck damage to fruit trees. *Ecological Applications* 1(1), 98-103.

Swihart, R. K. (1991b). Influence of *Marmota monax* on vegetation in hayfields. *Journal of Mammalogy* 72(4), 791-795.

Swihart, R. K. & Picone, P. M. (1994). Damage to apple trees associated with woodchuck burrows in orchards. *Journal of Wildlife Management* 58(2), 357-360.

Thomas, P. J., Mineau, P., Shore, R. F., Champoux, L., Martin, P. A., Wilson, L. K., Fitzgerald, G., & Elliott, J. E. (2011). Second generation anticoagulant rodenticides in predatory birds: Probabilistic characterization of toxic liver concentrations and implications for predatory bird populations in Canada. *Environment International* 37, 914-920.

Timm, R. M. (1994a). Norway rats. In S. E. Hygnstrom, R. M. Timm, & G. E. Larson, (Eds.), *Prevention and Control of Wildlife Damage*. (Pp. B-105 – B-120), Lincoln, Nebraska: University of Nebraska Cooperative Extension.

Timm, R. M. (1994b). House mice. In S. E. Hygnstrom, R. M. Timm, & G. E. Larson (Eds.), *Prevention and Control of Wildlife Damage*. (Pp. B-31 – B-46), Lincoln, Nebraska: University of Nebraska Cooperative Extension.

Tobin, M. E. & Sugihara, R. T. (1992). Abundance and habitat relationships of rats in Hawaiian sugarcane fields. *Journal of Wildlife Management* 56, 815-821.

Tomich, P. Q. (1986). *Mammals in Hawaii, 2nd Ed.* Honolulu, Hawaii: Bishop Museum Press. 375 pp.

Towns, D.R., Byrd, G.V., Jones, H.P., Rauzon, M.J., Russell, J.C. & Wilcox, C. (2011). Impacts of introduced predators on seabirds. In C. P. H. Mulder, W. B. Anderson, D. R. Towns, & P. J. Bellingham (Eds.), *Seabird Islands: Ecology, Invasion, and Restoration*. (Pp. 56-90), New York, New York: Oxford University Press.

Towns, D. R., Wardle, D. A., Mulder, C. P. H., Yeates, G. W., Fitzgerald, B. M., Parrish, G. R., Bellingham, P. J., & Bonner, K. I. (2009). Predation of seabirds by invasive rats: multiple indirect consequences for invertebrate communities. *Oikos* 118, 420-430.

United States Department of the Interior Fish and Wildlife Service. (2010). *Environmental Assessment – Invasive species eradication for habitat restoration on Tangik, Poa, and Sud Islands, Alaska*. Unpublished Report of USDI Fish and Wildlife Service, Homer, Alaska. 46 pp.

Veitch, C. & Clout, M. (Eds.). (2002). Turning the tide: the eradication of invasive species. *IUCN SSC Invasive Species Specialist Group*. Gland, Switzerland and Cambridge, UK.: IUCN.

Wilson, D.E. & Reeder, D.M. (2005). *Mammal Species of the World*. 3rd Ed. Baltimore, Maryland: Johns Hopkins University Press. 2,142 pp.

Winn, J. (Ed.). (1986). *Identifying and Managing Aquatic rodents in Texas: Beaver, Nutria and Muskrats*. College Station, Texas: Texas Agricultural Extension Service. 46 pp.

Witmer, G. W. (2004). Rodent ecology and plague in North America. In: *Proceedings of the 19th International Congress of Zoology*. (Pp. 154-156), Beijing, China: China Zoological Society.

Witmer, G. (2007). The ecology of vertebrate pests and integrated pest management (IPM). In M. Kogan & P. Jepson (Eds.), *Perspectives in Ecological Theory and Integrated Pest Management*. (Pp. 393-410), Cambridge, UK: Cambridge University Press.

Witmer, G., Boyd, F., & Campbell, E. (1998). Rat management for endangered species protection in the U.S. Virgin Islands. In R. Baker & C. Crabb (Eds.), *Proceedings of the 18th Vertebrate Pest Conference*. (Pp. 281-286), Davis, California: University of California.

Witmer, G. W., Boyd, F., & Hillis-Starr, Z. (2007a). The successful eradication of introduced roof rats (*Rattus rattus*) from Buck Island using diphacinone, followed by an irruption of house mice (*Mus musculus*). *Wildlife Research* 34, 108-115.

Witmer, G. W., Burke, P. W., Jojola, S., & Nolte, D. L. (2008). A live trap model and field trial of a nutria multiple capture trap. *Mammalia* 72, 352-354.

Witmer, G. & Eisemann, J. D. (2007). Rodenticide Use in Rodent Management in the United States: An Overview. In D. L. Nolte, W. M. Arjo, & D. H. Stalman (Eds.), *Proceedings of the 12th Wildlife Damage Management Conference.* (Pp. 114-118), Corpus Christi, Texas: 12th Wildlife Damage Management Conference.

Witmer, G., Eisemann, J., & Howald, G. (2007b). The use of rodenticides for conservation efforts. In D. L. Nolte, W. M. Arjo, & D. H. Stalman (Eds.), *Proceedings of the 12th Wildlife Damage Management Conference.* (Pp. 160-166), Corpus Christi, Texas: 12th Wildlife Damage Management Conference.

Witmer, G. W., Fall, M., & Fiedler, L. (1995). Rodent control, research, and technology transfer. In J. Bissonette & P. Krausman (Eds.), *Integrating people and wildlife for a sustainable future: Proceedings of the First International Wildlife Management Congress.* (Pp. 693-697), Bethesda, Maryland: The Wildlife Society.

Witmer, G. W. & Fuller, P. L. (2011). Vertebrate species introductions in the United States and its territories. *Current Zoology* 57(5), 559-567.

Witmer, G.W. & Hall, P. (2011). Attempting to eradicate invasive Gambian giant pouched rats (*Cricetomys gambianus*) in the United States: Lessons learned. In C. Veitch, M. Clout & D. Towns (Eds.), *Island Invasives: Eradication and Management.* (Pp. 131-134), Gland, Switzerland: International Union for Conservation of Nature (IUCN).

Witmer, G. & Jojola, S. (2006). What's up with house mice? – A review. In R. M. Timm & J. M. O'Brien (Eds.), *Proceedings of the 22nd Vertebrate Pest Conference.* (Pp.124-130), Davis, California: University of California.

Witmer, G.W., Pierce, J. & Pitt, W.C. (2011). Eradication of invasive rodents on islands of the United States. In C. Veitch, M. Clout & D. Towns (Eds.), *Island Invasives: Eradication and Management.* (Pp. 135-138), Gland, Switzerland: International Union for Conservation of Nature (IUCN).

Witmer, G. & Singleton, G. (2011). Sustained Agriculture: The Need to Manage Rodent Damage. In F. C. Wager (Ed.), *Agricultural Production.* (Pp. 1-38), New York, New York: Nova Science Publishers, Inc.

Witmer, G., Snow, N., & Burke, P. (2010). Potential attractants for detecting and removing invading Gambian giant pouched rats (*Cricetomys gambianus*). *Pest Management Science* 66, 412-416.

Yensen, E. & Sherman, P. W. (2003). Ground Squirrels. In G. A. Feldhamer, B. C. Thompson, & J. A. Chapman (Eds.), *Wild Mammals of North America.* (Pp. 211-231), Baltimore and London: The Johns Hopkins University Press.

In: Invasive Species
Editors: Joaquín J. Blanco and Adrian T. Fernandes

ISBN: 978-1-61942-761-7

Chapter 3

A Multi-Disciplinary Approach for Developing Tools to Monitor Invasive Buprestid Beetle Species

Michael J. Domingue and Thomas C. Baker
The Pennsylvania State University, Department of Entomology,
Center for Chemical Ecology, University Park, PA, US

Abstract

Buprestid beetles (Coleoptera: Buprestidae), which were previously considered to be minor forest pests, have received substantial attention recently due to the widespread ash tree mortality associated with the North America introduction of the emerald ash borer, *Agrilus planipennis* from Asia. At the same time, geographic expansions of forest buprestid species associated with other hosts have been noticed, causing concern about potentially similar destructive outcomes. Attempts have been made to understand the host-finding and mating behaviors of these forest buprestids in order to develop detection and management tools, with a focus primarily on *A. planipennis*. A complex sequence of chemical and visual signals appears to mediate such behaviors, which has been difficult to mimic in an effective trapping approach. Both sexes appear to be attracted to the tree canopy as a result of host-produced volatile compounds. Once in the canopy, males visually seek and fly directly onto stationary females from a distance of up to 1 meter, an event which immediately precedes attempts to copulate. The subsequent copulation attempts by the males are mediated by sex-specific cuticular hydrocarbon signals. Similar visually-mediated approaches toward dead models have also been observed in other tree-dwelling buprestids such as *Agrilus cyanescens*, *A. subcinctus*, *A. biguttattus*, *A. sulcicollis*, and *A. angustulus*. While the extent of similarities of both visual and chemical signals shared by these species and *A. planipennis* has not been fully explored, it is becoming clear that the visual signal that induces attraction to a female is broadly tuned, exhibiting substantial cross-species attraction. Despite subtle differences in coloration patterns, all of these species generate an intense spectral signal using the refractive properties of repetitive layering of the elytral cuticle. Further investigation into both the means of generation and response to the visual signal, as well as the chemical ecology of these species, may provide advances in trapping technologies for the detection of all tree-feeding Buprestids.

1. INTRODUCTION

The eruptive outbreak of the emerald ash borerer, *Agrilus planipennis*, in North America over the past decade has resulted in a serious threat to urban and forest ecosystems. The continuing spread of this insect has become an important model for understanding the ecology of invasive species. Furthermore, it has also highlighted the generalized potential threat of buprestid beetles (Coleoptera: Buprestidae), which like other invasive species have experienced host-range shifts because of human activities. Such impacts range from accidental introductions associated with travel or trade to climate-change-induced shifts in host susceptibility. The Buprestidae is a large and diverse family of insects, estimated to include 15,000 thousand species (Bellamy, 2008). Buprestids are typically associated with trees and shrubs, upon which eggs are laid in bark crevices. Larvae then tunnel into the inner bark where they feed to maturity and pupate before emerging from characteristic D-shaped exit holes as adults. The adults disperse and locate new hosts. In many cases they are known to intermittently feed and mate on the foliage of associated host plants. While this life history pattern is overwhelmingly common in buprestids, there are exceptions, such as leaf-mining in the genus *Brachys* (Turnbow and Franklin, 1981). Adult buprestids are also known for their bright, metallic coloring, which has long been implicated as a mechanism for mate attraction. Beer bottles having similar coloration and indentations as female *Julodimorpha bakewelli* were shown to evoke male mounting and copulation attempts (Douglas, 1980; Gwynn and Rentz, 1983).

The emergence of *A. planipennis* as a severe pest in North America has led to a renewed interest in research into the behavior and ecology of buprestids with the goal of developing methods for early detection of their infestations. Traditionally the impact of most buprestid populations has been assessed by examining damage done to host plants, which can be accomplished by observing the appearance of exit holes, branch dieback, or the presence of larval galleries under debarked logs and branches. The *A. planipennis* introduction to North America has provided a unique occasion in which a buprestid species has spread while infesting large numbers of healthy host trees within its host population, causing nearly universal mortality shortly after the appearance of physical symptoms of decline. In such a situation, detection of incipient infestations of healthy forests is crucial in order to maximize management options. Because there was no precedent for such a severe problem involving a buprestid species, little was previously known about the biology and ecology of any members of this group, let alone the emerald ash borer.

After approximately a decade since the initial identification of *A. planipennis* in North America (Haack et al., 2002), it is useful to review the progress of research and technological development aimed at its detection, while also identifying how such insights can be transferred to managing similar potential threats from other buprestid species. Here we will first present an overview of the buprestid species of greatest economic concern, and how their geographic ranges may be spreading. We will then focus on recent advances in knowledge of olfactory and visual mechanisms of host location and mating behavior in buprestids, with an emphasis on the well-studied *A. planipennis* model system. Finally, an assessment will be made of how current detection technologies are able to exploit such behavioral attributes of buprestids, with an exploration of how continued behavioral research will likely contribute to the development of improved traps.

2. BUPRESTIDS OF ECONOMIC CONCERN

a) The Emerald Ash Borer, *Agrilus Planipennis* Fairmaire

The emerald ash borer is a buprestid species native to North Eastern Asia (Jendek, 1994). It feeds primarily on ash trees (*Fraxinus* spp.), both in its native range (Yu, 1992), and in North America (Anulewicz et al., 2008).

The biology is typical of that described for other tree-feeding buprestids (Yu, 1992; Cappaert et al., 2005; Wang et al., 2010) The first positive identification of emerald ash borer in North America occurred near Detroit, Michigan, USA and Windsor, Ontario, Canada in 2002 (Haack et al., 2002). Soon after this identification, the beetle was found in other nearby Great Lakes states (Poland and McCullough, 2006). The beetle has now been detected in a much broader geographical range including Quebec, Canada and many northern U.S. states, which span the Mississippi River to Atlantic Coast regions, where the infestation is expected to continue to grow (MacFarlane and Meyer, 2005; Kovacs et al., 2010, 2011). Throughout its known range in North America, there have been widespread reports of overwhelming mortality of ash trees (Poland and McCullough, 2006).

It is believed that dispersal of *A. planipennis* has been assisted by the movement of infested firewood (Muirhead et al., 2006), in which the beetles are viable given typical handling procedures (Petrice and Haack, 2006). Quarantine efforts have thus focused on limiting the movement of firewood away from areas of known infestations. Within quarantine zones it has been estimated that infestations have spread at a rate of only 1km per year (Sargent et al., 2006), further underscoring the role of human assistance in the dynamics of the more widespread multi-state outbreak of this pest.

Figure 1. The five buprestid species used as decoys for attracting European oak buprestids *Agrilus planipennis* (8–15 mm), *A. biguttatus* (10–14 mm), *A. sulcicollis* (6–8.5 mm), *A. cyanescens* (4.5–7.5 mm), and *A. angustulus* (4–6.5 mm). List includes range of lengths (Muskovits and Hegyessy, 2002). Reproduced from Domingue et al., 2011.

b) The Bronze Birch Borer, *Agrilus anxius* Gory

The bronze birch borer is a pest of trees in the genus *Betula* in North America. Normally only stressed or diseased trees are seriously affected, as a result of better larval performance in these conditions (Ball and Simmons, 1986). *A. anxious*-related birch decline has been shown to be correlated with stresses from temperature and precipitation changes (Jones et al., 1993). There is also strong evidence that susceptibility to this pest differs among birch species, with exotic tree species experiencing greater infestation, damage, and mortality rates than native species (Miller et al., 1991; Nielsen et al., 2011).

c) North American Oak Borers

The gold spotted oak borer was originally considered to include the single species, *A. coxalis* Waterhouse, which has recently been separated into sister species, *Agrilus auroguttatus* Schaeffer and *A. coxalis* (Hespenheide and Bellamy, 2009). The borers are native to the southwestern United States and Central America, feeding on oak trees within that range. Severe infestations of *A. auroguttatus* have been implicated in widespread tree mortality of coast live oak in California, after a likely recent migration of the species from Arizona (Coleman and Seybold, 2008, 2011). The susceptibility of coast live oak to attack and mortality from this species is not related to stress conditions, underscoring the likely severity of this impending threat (Coleman et al., 2011).

The two-lined chesnut borer, *Agrilus bilineatus* (Weber) is another pest of oak that is found in the eastern United States and Canada. The biology is similar to that of other tree-feeding buprestids described here (Cote and Allen, 1982), with attacks having been noticed to be prevalent on stressed trees, beginning high in the upper branches, and proceeding downward until the tree is killed (Haack and Benjamin, 1982). The prevalence of two-lined chesnut borers and associated oak tree mortality has been linked to defoliation events caused by the Gypsy moth, *Lymantria dispar* (Muzika et al., 2000)

d) European Oak Borers

Among several buprestid species known to feed on European oaks (Muskovits and Hegyessy, 2002), *Agrilus biguttatus* Fabricius (Figure 1B) is considered to be the greatest threat. It will sometimes attack larger diameter stems and branches, causing oak decline and mortality (Moraal and Hilszczanski, 2000; Vansteenkiste et al., 2005). Warm, dry summer weather encourages its growth and reproduction (Schlag, 1992). Mass occurrences of this species are often recorded 1–2 years after outbreaks of defoliating insects such as *Lymantria dispar*, causing further decline of weakened trees (Csóka and Kovács, 1999; Csóka and Hirka, 2006; McManus and Csóka, 2007).

Agrilus sulcicollis Lacordaire (Figure 1C), and *A. angustulus* Illiger (Figure 1E), are examples of smaller species not as strongly associated with oak decline (Schopf, 1992). They attack smaller branches and the upper parts of stems and smaller trees (Kolk and Starzyk, 1996; König, 1996; Csóka and Kovács, 1999; Muskovits and Hegyessy, 2002). *A. sulcicollis*, has recently been identified in North America (Haack et al., 2009; Jendek and Grebennikov,

2009), prompting concern that there may be a pathway by which the more aggressive *A. biguttatus* might also reach the forests of North America.

e) Other Forest Systems Affected by Buprestids

Although the above examples provide some of the more serious pest management concerns regarding buprestids, there are diverse buprestids common throughout all known forest ecosystems. When other species have emerged as threats of interest, the potential deleterious impacts have always included either drought stress of trees, or the accidental exotic introduction of the buprestid species. Outbreaks of the beach splendor beetle, *Agrilus viridis* Linnaeus, were reported in Germany corresponding to a severe drought in 2003 (Lakatos and Molnár, 2009). Likewise, *A. difficilis* Gory, a known buprestid pest of honeylocust in North America, is more successful on drought-stressed trees (Wescott, 1973; Akers et al., 1986). *Agrilus subrobustus* Saunders is an Asian buprestid species that has recently been found in the southeastern United States (Hansen et al., 2010; Hoebeke et al., 2011). It is a pest of Mimosa or silktree, *Albizia julibrissin*, which itself was originally imported to North America approximately 200 years ago.

f) Agricultural Pests

There are also buprestids of economic concern that are pests of fruit-bearing trees and shrubs in plantation settings. In such circumstances mortality is not necessary to cause significant damage to the crops produced by affected plants. The flatheaded apple tree borer, *Chrysobothris femorata* Olivier, a widespread species, and a related species, the pacific coast borer, *Chrysobothris mali* Horn, are sources of such damage for several fruit and nut trees in North America (Burke, 1919). Particularly vulnerable to such attacks are newly transplanted trees (Johnson and Lyon, 1991). A number of other buprestids are known to affect berry plants, often causing galls in the woody stems. For example, the rednecked cane borer, *Agrilus ruficollis*, causes galls in blackberry bushes that have been linked to reduced berry counts (Johnson and Mayes, 1989; Johnson, 1992) *Agrilus rubicola* is a species native to Europe that has long been noted as a pest of agricultural crops such as raspberries (Davis and Raghuvir, 1964).

3. Exploitable Behaviors of Buprestids: An Ecological Perspective

The species listed above clearly share many biological attributes, such as similar life history patterns and a tendency to cause greater damage to drought-stressed or non-native trees. However, little is known about several of the critical behavioral attributes of buprestids that might be exploited for management purposes, especially the development of detection and monitoring tools. Detection tools are perhaps most critical for identifying introductions of exotic buprestids, which are unpredictable events that may go unnoticed until substantial tree damage and mortality has already occurred.

Trapping approaches have usually targeted the reproductive behavior of insects. After the first insect sex pheromone was identified (Butenandt et al., 1959) the discovery of pheromones of agricultural and forest pests soon began to intensify (Silverstein et al., 1966; Roelofs and Arn, 1968; Roelofs et al., 1969; Tumlinson et al., 1969; Bierl et al., 1970), leading to the discovery of a dizzying array of chemical signaling pathways in a diverse number of insect orders and families. The identification of such insect pheromones has since made an overwhelmingly large contribution to the development of pest management tools, which have come to include detection, mating disruption, and mass trapping (recently reviewed, Baker, 2009).

Scolytid bark beetles fill a quite similar ecological niche to the buprestid beetles described above, with both groups feeding on the inner bark of trees. Scolytids were among the first insects for which pheromones were identified (Rudinsky, 1966; Silverstein et al., 1966, 1968; Pitman et al., 1968; Wood et al., 1968). From the family Scolytidae, the website pherobase.com lists the pheromones of 72 species from 21 genera, most of which involve long-range attraction. However, there are currently no known long-range pheromones for any buprestid species. Here we propose that this lack of reporting of buprestid pheromones is not due to a failure of chemical ecologists to isolate them, but rather is more likely to reflect their lack of existence. Despite the similar use of the inner bark as the nutritive source for larval development in scolytids and buprestids, differences in particular behavioral and life history traits make long-range pheromone use advantageous for scolytids, but disadvantageous for buprestids.

The advantages to each individual scolytid bark beetle in emitting a pheromone signal that will attract an aggregation of conspecifics have been well described (Berryman et al., 1985). Scolytid adults enter the bark after dispersing. They then emit their pheromones from under the bark, where mating subsequently occurs, followed by female excavation of galleries for laying eggs. It is advantageous for the survival of each individual adult bark beetle attempting to enter the bark to broadcast an aggregation pheromone, because they are otherwise likely to be overwhelmed by the tree defenses if they are not joined by large numbers of other beetles (Raffa and Berryman, 1983). Furthermore, the bark provides some protection from predators and parasitoids, which are known to eavesdrop on pheromone signals (Bakke et al., 1981; Billings et al., 1984).

In contrast, adult buprestids do not enter the bark. Instead, each female beetle potentially lays a multitude of eggs on the bark, which will emerge as larvae that will later attempt to enter the bark and overwhelm any defenses produced by the tree. Once a suitable host has been found, there is a substantial disadvantage in broadcasting a chemical signal that intraspecific or interspecific competitors or predators might be able to exploit. In buprestids, there thus should be a strong selective pressure in females to develop the ability to locate favorable oviposition sites in stressed trees (Carlson and Knight, 1969). For males, finding suitable hosts would also be advantageous if it helps them locate females. In such a scenario it is not requisite for mating to occur directly at the oviposition sites on preferred hosts. Indeed female A. planipennis appear to be most commonly approached for mating when they are feeding on leaves (Lelito et al., 2007), indicating a spatial separation of ovipostion and mating behavior. Thus the location of females by males after they have located such preferred hosts may be assisted by other intermediate to short range chemical or visual cues.

As in many pest management scenarios, fully exploiting all possible dispersal and mate-searching behaviors is needed to develop a trap that most effectively functions at very low population densities. Regardless of the reproductive capability of the target species, such low population densities will exist at the earliest stages and advancing margins of outbreaks of invasive species. Given the context of the life history characteristics of buprestids described above, it is clear that exploiting male mating behavior will be most effective for trapping. One reason males should be easier to detect at low densities is that unlike females, they are expected to continue to move about in the environment searching for mates among different favorable host trees throughout their adult lifespan. Females, on the other hand, are likely to become less mobile once favorable oviposition sites have been located. Secondly, male mating behavior has a point-source orientation towards its target of a receptive female. Female host selection has an inherently weaker directional component, with orientation to stress-induced volatiles at a much larger spatial scale (the whole tree). Furthermore, it is not necessarily clear that stress volatiles used by females to locate trees will always emanate from the tissues where oviposition occurs. Thus, as we review the literature that is now available on host location and mating behavior in buprestids, we will focus on the development of technology for detection of males.

4. The Role of Stress-Induced Volatiles in Host Location

With the observation that tree stress is associated with buprestid infestations, experiments have been performed to attempt to increase attraction to trees by manipulating stress levels and volatile production of trees likely to be affected by buprestid pests. For *A. bilineatus*, approaches such as artificiallygirdling trees (Dunn et al., 1986a; Haack et al., 1982) or adding cut logs and ethanol (Dunn and Potter, 1991) have proved effective in attracting adult beetles. Furthermore, steam distillates from stressed oak logs were also used to attract more *A. bilineatus* to traps (Dunn et al., 1986b). A similar and more detailed exploration has been undertaken of the attraction of *A. planipennis* to stress-induced volatiles, which has been demonstrated to occur after girdling of ash trees (McCullough et al., 2009). Crook et al. (2008) identified several ash bark sesquiterpenes that were also antennally active to male or female *A. planipennis.* All but one of these compounds are found in commercially available manuka oil, a steam distillate of the manuka tea tree *Leptospermum scoparium*. All of the desired compounds are found in phoebe oil from the Brazilian walnut, *Phoebe porosa* (Crook et al., 2008; Cossé et al., 2008). Crook et al. (2008) were able to capture *A. planipennis* at an elevated rate when manuka oil or phoebe oil was added to traps at a variety of doses.

These oils have proven most effective when presented with stand-alone traps. For example, when manuka oil was added directly to trees, capture of *A. planipennis* was not elevated over control trees and was lower than in girdled trees (McCullough et al., 2009). However, when purple prism-panel traps were use near ungirdled trees with phoebe or manuka oil being added, trap captures were better than if similar unbaited traps were hung near girdled trees (Marshall et al., 2010).

Although sex ratio varies significantly within different trapping events in different studies, there has been no evidence that manuka or phoebe oil affects sex ratio. Studies employing manuka oil have shown enhancement of trap capture on purple prism traps versus controls without altering sex ratios (Grant et al., 2010; Silk et al., 2011; Lelito, 2009). Phoebe oil added to green prism traps similarly increases trap captures versus controls without substantially changing the sex ratio (Grant et al., 2011; Lelito, 2009).

5. Leaf- and Beetle- Produced Compounds Released When Adults Feed

The observation of feeding by adult buprestids on the leaves is common, and for female *A. planipennis* precedes mating (Lelito et al., 2007). Thus this behavior provides the possibility for both the induction of host volatiles and the release of sex pheromones. It has been found that both *A. planipennis* feeding and the application of methyl jasmonate to Manchurian ash seedlings induce similar increases in volatile emissions from leaves, but with a different qualitative profile (Rodriguez-Saona et al., 2006). Several of these volatile compounds were also found to be antennally active to males and females by GC-EAD analysis (Rodriguez-Saona et al., 2006; deGroot et al., 2008). Initial lab bioassays indicated that only females were behaviorally active in response to such compounds (Rodriguez-Saona et al., 2006). However, subsequent testing of these compounds in the field revealed a strong male bias in trap capture increases in response to (*Z*)-3-hexenol, the only consistently effective green leaf volatile (deGroot et al., 2008; Grant et al., 2010, 2011). The discrepancy between lab and field behavior may be explained by the subsequent field observations that males are always flying in search of mates when they are in a position where they would be exposed to such volatile compounds (Lelito et al., 2007, 2011; Domingue et al., 2011), whereas the laboratory bioassays involved walking approaches.

A possible sex pheromone was inferred from a study involving *A. bilineatus* where males were attracted to caged females provided with leaves for feeding (Dunn and Potter, 1988). This attraction occurred regardless of whether or not cut logs were included with the females. It was not clear if other visual or auditory signals could have also caused this attraction of males. In laboratory experiments, *A. planipennis* were able to locate females at close range when their eyes had been blocked by paint, but not when their antennae were blocked (Pureswaran and Poland, 2009). Other natural environmental cues necessary to elicit visual mate finding were absent in this experiment, such as direct sunlight and host leaves, likely precluding the observation of visual-mate finding when the eyes of males were uncovered. Thus, while the relative importance of visual versus chemical attraction could not be assessed in this experiment, there was nevertheless likely a cue perceived by the antennae at short-range that promoted mating.

The nature of the chemical cue operating in this experiment by Pureswaran and Poland was not determined, but a prospective volatile pheromone for *A. planipennis* has been isolated from females feeding from ash leaves (Bartelt et al., 2007). The compound was (3*Z*)-dodecen-12-olide, a macrocyclic lactone that was shown to evoke antennal responses in both males and females (Bartelt et al., 2007). It was later discovered that exposure to sunlight causes isomerization of this compound to (3*E*)-dodecen-12-olide, which is also antennally active (Silk et al., 2011). Field trapping has shown that both of these isomers can enhance the capture of male *A. planipennis,* but only when acting synergistically with (*Z*)-3-hexenol.

Thus, while further research is needed to determine the optimal release rates for all possible effective green leaf volatiles and pheromone components, there is now good evidence that female *A. planipennis* provide males with an exploitable attractive chemical cue while they are feeding on the foliage.

6. Visual Mate-Finding

Although intermediate-to-short-range semiochemical signals produced by feeding females may play a role in male mate-finding, there is substantial evidence that such signals are not necessary to elicit mating approaches by males, which can occur solely due to visual signals. Using dead pinned *Agrilus planipennis* beetles as models on ash leaves, Lelito et al. (2007) were able to demonstrate highly repeatable male flights directly onto the decoys from 30 -100 cm above (Figure 2). The term "paratrooper copulation" was used to describe this behavior as it was often immediately followed by vigorous attempts to copulate with the model, which involved the extension of the feral male's aedeagus.

It was quite clear that the visual image of the beetle alone is capable of evoking this response in the field, because the rate of male paratrooper flights was nearly identical if either male or female models were used and regardless of whether or not all potential chemical cues had been washed off of the models with organic solvents (Figure 3, Lelito et al., 2007). Also determined to be important was the wing position of the models approached. To induce male mating flights, the wings of the model needed to be held flatly, fully covering the abdomen.

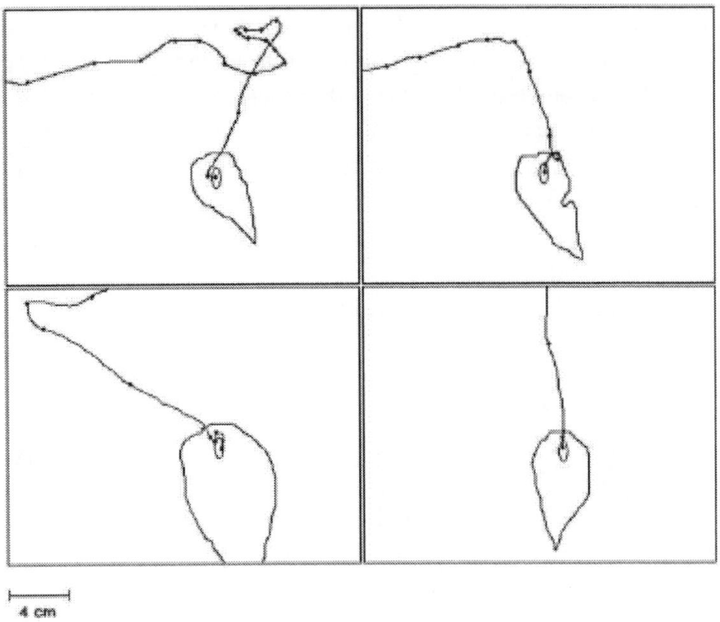

Figure 2. Various approach flight-tracks, transcribed from video, of feral male *Agrilus planipennis* descending rapidly onto pinned, closed-elytra decoys on leaves. The scale bar at left indicates 4 cm in each frame. The small oval on each depicted leaflet is the position of the pinned *A. planipennis* on the leaf. The span of time between each dot is 1/16 s. Reproduced from Lelito et al., 2007.

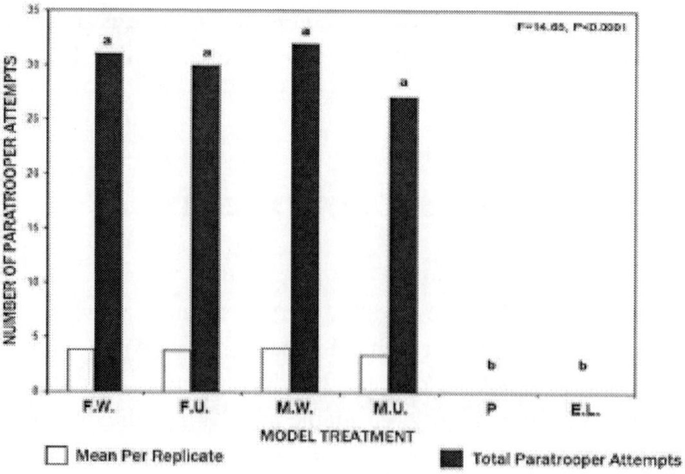

Figure 3. Mean number of *Agrilus planipennis* paratrooper copulation attempts observed per replicate (black bars) and the total number of paratrooper attempts (empty bars) observed in response to the different models during the eight replicates of our experiment. F.W. washed female; F.U. unwashed female; M.W. washed male; M.U. unwashed male; P leaflet with pin; E.L. empty leaflet. Eight replicates of this experiment were conducted, each lasting for 2 h. The number of feral males observed near any of the four types of beetle models is not statistically different (ANOVA, $F=0.47$, $p=0.7024$) but all four of the decoys had significantly more males than either of the two controls (ANOVA, $F=14.65$, $p<0.0001$). Reproduced from Lelito et al., 2007.

Spreading the wings of the models to reveal the abdomen reduced the rate of these approaches. Additional experiments determined that the use of only two excised elytra placed in parallel could also induce the paratrooper approaches at a high frequency (Lelito, 2009).

Subsequently, similar behaviors have been documented in other tree-dwelling buprestids when field populations have been presented with such visual decoys. Both *Agrilus cyanescens* Ratzeburg and *A. subcinctus* Gory were observed performing similar paratrooper flights toward pinned conspecific models (Lelito et al., 2011). *A. cyanescens* is a native European species that is now found in North America and specializes on honeysuckle (*Lonicera*) host plants. *A. subcinctus* is a native North American species that feeds on ash trees. Figure 1D shows the relatively smaller size of *A. cyanescens* in comparison to *A. planipennis*. *A. subcinctus* are even smaller, approximately similar in length to the *A. angustulus* pictured in Figure 1E. As was previously determined for *A. planipennis*, washing with solvent did not affect the frequency of aerial approaches to decoys for either species.

Attempts to assess the species-specificity of male aerial approaches to visual decoys have revealed a surprisingly degree of promiscuity. When examining the mating behavior of *A. cyanescens* it was found that the iridescent elytra of tiger beetles (*Cicindela*) could elicit male approaches at a frequency as great as conspecific decoys (Lelito et al., 2011; Figure 4). Field observations of the three sympatric species of European oak buprestids, *A. biguttatus*, *A. sulcicollis*, and *A. angustulus*, again revealed visually mediated mating approaches from distances up to 1m above pinned decoys (Domingue et al., 2011). Furthermore, in this experiment the feral beetles observed were presented with the choice of decoys of five different species (all of those shown in Figure 1). *A. biguttatus*, *A. sulcicollis*, and *A.*

angustulus all responded to the decoys presented with a great deal of cross-species attraction, despite some indications of preferences for certain decoys (Figure 5).

Figure 4. A live male *Agrilus cyanescens* attempting copulation with a *Cicindela sexguttata* elytron treated with female *A. cyanescens* cuticular extract, lateral (A) and dorsal view (B). Reproduced from Lelito et al., 2011.

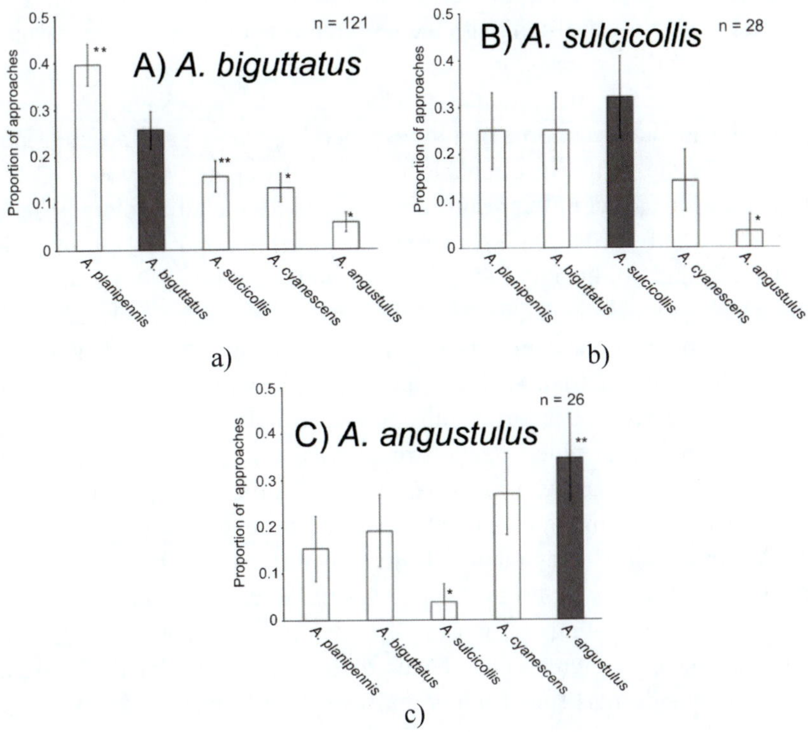

Figure 5. Mean (± SE) proportion of paratrooper approaches to each of the five decoys, by A) *Agrilus biguttatus*, B) *A. sulcicollis,* and C) *A. angustulus*. Significantly different proportion of paratrooper approaches in comparison to the conspecific decoy indicated (Fisher's exact test: * P<0.05, **P<0.1). Reproduced from Domingue et al., 2011.

For example, *A. biguttatus* actually preferred the larger *A. planipennis* decoys to conspecific decoys, while they approached the other smaller species, *A. sulcicollis*, *A. cyanescens*, and *A. angustulus* less often. Both *A. sulcicollis*, and *A. angustulus* showed similar attraction to the other decoys presented, with the exception of reduced attraction to each other.

Another pattern emerging from these studies is the observation that the smallest species, *A. subcinctus*, and *A. angustulus*, while dropping downward toward decoys from 1m above them like the larger species, did not usually land directly on their targets. Instead they tended to land 1-2 cm away and then approached the decoys, often antennating them (Lelito et al., 2011; Domingue et al., 2011). Perhaps the smaller beetles simply experience a more turbulent path dropping through the air toward their targets than the larger beetles.

Considering the prevalence of paratrooper approaches in buprestid males, the bizarre spectacle of buprestids attempting to mate with beer bottle pieces observed by Gwynn and Rentz (1983) was perhaps reflective of a universal inherent tendency of particular visual cues to induce the release of a strong mating response. These observations suggest a possible common visual template for male mate-attraction, which would be highly desirable to better understand for the further development of more effective and sensitive traps. Such a template would not preclude the ability for both the emitter and receiver of the visual signal to modify characteristics and preferences so as to allow certain species-specific information to be conveyed. Perhaps elytral spots, lines, or subtle color variations might provide information that modifies the response. One conceivable need for a modification of the general visual template might be to facilitate the avoidance of heterospecific mating with a competing sympatric species, which is consistent with the preference patterns observed for the distinctly colored *A. sulcicollis* and *A. angustulus*.

A further understanding of both the production and response to the visual signals produced by buprestids is needed to begin to test such hypotheses and develop the technical capacity to duplicate the signal in a reliable fashion for trapping purposes. Currently it is known that the metallic aspect of buprestids and many other metallic-colored beetles such as cicindelids, is structurally produced by alternating uncolored and melanized cuticle layers (Figure 6) (Durrer and Villinger, 1972; Schultz and Rankin, 1985). This one quarter wavelength-stacked structure is highly effective for creating a strongly spectral color in sunlight, with little background emission of light of other wavelengths. Furthermore, the wavelength of light reflected from such a structure may vary with the angle of incidence, creating an iridescent effect. Such structurally produced colors are also often known to be accompanied by a dimpled or pitted coarse structure, which can be microscopically viewed for Buprestids (Figure 7). It has been suggested that such structures can obscure the different spectral colors emanating from the cuticle when viewed from a distance by predators (Seago et al., 2009). A more detailed examination of the structural coloration produced by a wider array of buprestid beetles may be useful for understanding which aspects of their coloration elicit the intense male mating responses that have been observed. Ultimately the nanofabrication of such structures could be used to create artificial beetle decoys to be incorporated in trapping designs aimed at more fully exploiting male mating behavior.

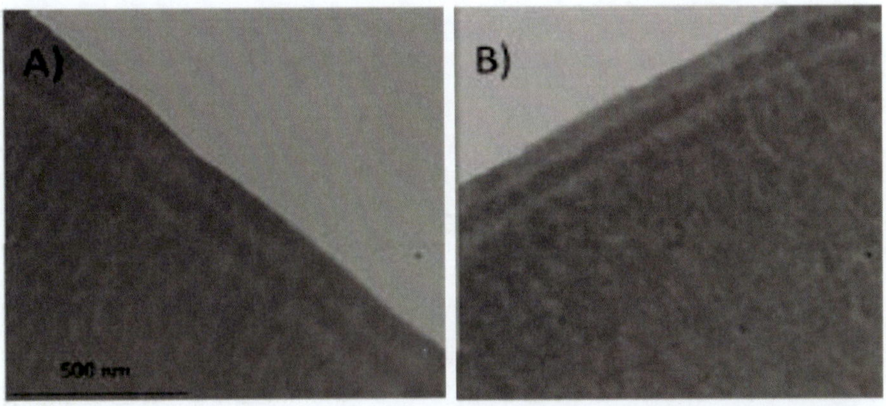

Figure 6. TEM image of the elytra of *A. biguttatus* (left) and *A. sulcicollis* (right) showing regular layering of the epicuticle.

Figure 7. *Agrilus sulcicollis* under a light microscope revealing irregularities of the elytral surface structure.

7. CONTACT PHEROMONES

After buprestid males have visually located and approached females, there are additional cues known to facilitate copulation and prolonged mating contact. When presenting solvent-washed or unwashed beetle decoys, Lelito et al. (2007) found that feral *A. planipennis* males would stay on top of unwashed females for much longer periods than on washed models of either sex or unwashed males. The cuticular hydrocarbons 3-methyl-tricosane (Lelito et al., 2009) and 9-methyl-pentacosane (Silk et al., 2009) were later implicated in causing this prolonged mating contact. Although no contact pheromone components have been identified for other buprestids, similar behavioral evidence was found for them in *A. cyanescens* and *A. subcinctus*, which both had greater durations of copulatory contact when presented with unwashed females versus other treatments (Lelito et al., 2011). *A. biguttatus* males spent more time on *A. planipennis* dead female decoys than on conspecifics, suggesting that if *A. biguttatus* does use a contact pheromone, it is likely similar to that of *A. planipennis* (Domingue et al., 2011). There have not yet been any attempts to employ contact pheromones in trapping designs.

8. Detection of Buprestids of Economic Concern

a) Visual Detection Methods

In many cases a high incidence of buprestid damage can be observed by various types of visual surveys of trees that might be expected to be infested by beetles. Such detection methods tend to be labor intensive and are usually most effective if used in an area where stressed or vulnerable trees exist. In some cases this approach has consisted simply of peeling back bark (Wargo, 1977). The characteristic D-shaped emergence holes of buprestids can also be used to estimate the presence and size of infestations. For example, binocular surveys of emergence holes at the edge of an outbreak of *A. planipennis* proved capable of predicting crown thinning rates (Smitley et al., 2008). Another highly effective technique for capturing buprestids is the use of trap trees. This technique involves girdling trees and allowing them to be colonized by the buprestids (McCullough et al., 2009; Marshall et al., 2011). Some studies have placed a sticky band around the trunk of a girdled tree to detect beetles (Dunn et al., 1986a). Girdling is highly effective for attracting buprestids, but is of course also destructive. Finally, it has been shown that early infestations of *A. planipennis* in at-risk areas can be detected in a high percentage of trees by removing and sampling 6 inch branches (Ryall et al., 2011). Such a technique provides a minimally destructive approach for identifying large numbers of individually infested trees for treatment or removal, before they can serve as sources of continuing population expansion.

The visual survey, girdling, and branch sampling techniques described above are most effective for urban situations where areas of infestation are known and it is of interest to characterize and simultaneously mitigate the extent of the problem. For a highly invasive species such as *A. planipennis* there is also a need to survey larger geographic areas to determine where new infestations are emerging, often in areas that can be quite distant from known infestations when human-assisted emigration has occurred. For such an application, traps are desirable that can be easily fabricated and distributed to monitor large geographic areas. The following sections discuss recent progress toward developing such traps for buprestids.

b) Sticky Traps

A number of different techniques have been used to attempt to capture adult buprestids using natural or artificial surfaces coated with sticky material that will entangle them. One such approach for trapping buprestids has been to employ sticky card surfaces (Wargo and Montgomery, 1983). For *A. planipennis* it was initially demonstrated that purple panel traps made of corrugated plastic were more effective in capturing beetles when compared to a large array of similar traps using other colors (Francese et al., 2005). To provide a large trapping surface, this basic trap design was later modified to include a three-panel "prism trap". This trap, with each side being approximately 35 X 60 cm (Francese et al., 2008), has subsequently become the trap used in most *A. planipennis* detection programs. Francese et al. (2008) also found that more beetles were caught if traps were placed higher in the canopy and near open, sunlit areas.

The purple version of the prism trap was developed in response to the operational success of the materials employed, rather than to any particular biological principles. Thus, attempts to understand the behavioral attributes of buprestids being exploited by these traps have occurred subsequent to their original development. Crook et al. (2009) performed electroretinogram recordings on *A. planipennis* to wavelengths of light within the visual spectrum. He found generally broad responsiveness to all wavelengths, with a modest increase in sensitivity to red wavelengths in females. Crook et al. (2009) went on to speculate that the preference for purple colored traps might be linked to this greater response to the higher wavelengths of visible light in females, and perhaps also indicative of attraction to the red abdomen of *A. planipennis*. Alternatively, the same authors later proposed that the attraction to purple traps might arise from attraction to host bark (Francese et al., 2010a). The second explanation seems more plausible given that greater proportions of females have been documented for beetles on purple traps (Francese et al., 2010a), and for beetles observed on ash tree trunks and branches (Lelito et al., 2007). Furthermore, revealing the red abdomen of visual decoys actually severely reduced the attraction of males to the decoys (Lelito et al., 2007). In addition there is no known behavioral attraction of females toward conspecifics, let alone specifically to the abdomen.

It was also later found that a green version of the prism traps could often be more effective than the purple traps. Additional effort has been made to identify the optimal wavelength and reflectance of green traps used (Francese et al., 2010a). It has also been found that green traps perform better the higher they are placed in the canopy (Francese et al., 2010b). As described above, researchers investigating the effects of the green leaf volatile (*Z*)-3-hexenol and the prospective lactone pheromone (3*Z*)-dodecen-12-olide, have had greater success when using these green prism traps high in the canopy (Grant et al., 2010, 2011; Silk et al., 2011). The male-biased sex ratio observed in these studies suggests that these traps may be evoking some aspects of male mate-finding, which involves persistent flight through the canopy while actively searching for prospective mates. These attributes include the green color of leaves and the chemical emanations of leaves being fed upon by females when they become accessible for mating approaches.

c) Multi-Funnel Traps

In an attempt to find an alternative to the large sticky prism traps, Francese et al. (2011) developed modified multi-funnel traps, based on existing designs developed for scolytid bark beetles (Lindgren, 1983). Such traps function by inducing beetles to land on their sloped surfaces where they slip and fall into a collection cup below. These traps were customized using the green and purple colors employed for prism traps. They were effective at catching similar numbers of *A. planipennis* as prism traps if coated with substances such as Rain-X or Teflon. These traps provide the potential advantage of avoiding the difficulty of working with the sticky glue material used for the prism traps. It is not yet known how effectively the semiochemical lures developed for *A. planipennis* work with this trap design. However, they seem to operate using similar principles as the prism trap, providing a large surface of a color that is presumably attractive to the beetles because of some biological function, such as oviposition or maturation feeding.

d) Visual Decoy-Based Traps

Agrilus planipennis can be trapped by applying Tanglefoot glue to leaves with pinned visual decoy beetles, an assemblage that was given the name "sticky leaf trap" (Figure 8A; Lelito et al., 2007, 2008). Further attempts to use *A. planipennis* decoys on plastic cards were also successful in capturing other *A. planipennis*, but not as successful as the natural leaf and decoy combination (Lelito et al., 2008; Lelito, 2009). It was also demonstrated that the use of manuka oil or phoebe oil can increase captures on such traps (Lelito, 2009). Whenever in the above-cited studies a visual lure was employed, and the sexes of beetles captured were identified, a heavy male bias was reported. Furthermore, leaves without decoys captured fewer beetles without displaying a male bias. Thus, it is clear that decoy-baited traps are positively exploiting the male visual mate-finding behavior described as "paratrooper" approaches.

It is impractical to use sticky leaves for sustained field trapping over large geographic areas due to the degradation of the leaf surface over time. More recently it was shown that green plastic cards were proficient at catching the European oak borers *A. biguttatus, A. sulcicollis*, and *A. angustulus* if presented on a horizontal surface located on branches to replace the foliage (Figure 8B; Domingue, 2011). Furthermore the addition of *A. planipennis* decoys further increased trap captures of European oak borers, as did concomitant baiting with manuka oil and *Z*-(3)-hexenol. To further improve usability of such traps, it would be desirable to develop an artificial decoy that is capable of being affixed to the traps before deployment.

e) Biological Detectors

It has been noted that a solitary digging wasp, *Cerceris fumipennis*, hunts buprestids, using them to provision their nests (Evans, 1971). A wide diversity of buprestids has been reported from samples of their nest sites (Marshall et al., 2005). Furthermore, researchers have been able to use these wasps to successfully capture *A. planipennis* near infestation sites (Careless and Marshall, 2010).

Figure 8. Visual decoy-based trapping designs: A) *Agrilus planipennis* on sticky leaf captures an *A. sulcicollis* male; B) *A. planipennis* on green plastic surface captures multiple small buprestids.

The potential for using *Cerceris fumipennis* as a detection tool is intriguing, and researchers are now implementing the monitoring of a large geographic area using a network of volunteers at research institutions (Careless and Marshall, 2010). Regardless of the success of such endeavors, we also believe that further research into the visual attributes of this wasp may eventually produce insights into how male buprestids themselves are able to find males. The wasps appear to be able to recognize universal characteristics of buprestids. As described above, the high degree of visual cross-attraction among buprestids suggests a common visual template for mate attraction. Thus, *Cerceris fumipennis* wasps may also be highly tuned to the essential visual attributes of buprestids that characterize this general template, which if duplicated may lead to a highly effective generic trap for buprestids.

CONCLUSION

We currently envision the further refinement of both the multi-funnel and the visual decoy-based trapping approaches. Unfortunately the two approaches are not likely to be compatible to employ within the same design. The surfaces of funnel traps slope downward and inward, and are thus inherently oriented away from sunlight. The presentation of decoys in full sunlight, however, appears to be a key component of visual-decoy-based trapping approaches. Nevertheless, the multi-funnel traps have the distinct advantage of allowing captures of large numbers of buprestids without entangling them in sticky material, thus allowing substantial improvement in the collecting, handling, and identification of insects found in these traps. Because they appear to operate using similar biological principles as prism traps, we envision that they will be optimized by expanding upon similar lines of research, and become the most convenient and effective tool for monitoring the sizes of buprestid populations in areas of known or likely infestations.

However, because long-range pheromone signaling does not seem to exist in buprestids, such traps may continue to experience difficulty attracting and capturing beetles from very small populations. The visual decoy trapping has a greater potential in this respect because it manipulates the strongest behavioral tendency yet observed in buprestids, the paratrooper landing of the highly vagile males onto prospective mates after their wide-ranging scanning of the tree canopies. These traps require further development for widespread usability, including the fabrication of artificial visually-attractive decoys. It would also be desirable to develop traps based on this principle that circumvent the use of sticky material. If these engineering challenges can be met, visual-decoy traps are likely to be well suited for distribution across large geographic areas for early detection of unexpected populations of buprestids.

Given that the host location, pheromone signaling, and visual mate-finding behaviors of *A. planipennis* potentially contributing to optimal trap design are not yet fully understood, applying such trap technologies to other buprestid species will also be another substantial challenge. All of the less thoroughly researched buprestids have tended to show similar biological properties and ecological interactions as those observed for *A. planipennis*. However, most of the details of these interactions are not yet known. For example, will most species use similar chemical cues when responding to stressed hosts? Do additional intermediate-range pheromones similar to (3*Z*)-dodecen-12-olide exist in other species? If so, are they relatively more or less important in comparison to the visual mating response of males in those cases? Ultimately it will be desirable to develop multifunnel and visual-decoy traps that will work generically for many buprestids species, but that can also be tuned to focus on particular species of interest when needed.

References

Akers, R. C., Herms, D. A. and Nielsen, D. G. (1986). Emergence and adult biology of Agrilus difficilis (Coleoptera, Buprestidae), a pest of honeylocust, *Gleditsia triacanthos*. *Great Lakes Entomol*. 19(1): 27–30.

Anulewicz, A. C., McCullough, D. G., Cappaert, D. L., and Poland, T. M. (2008). Host range of the emerald ash borer (Agrilus planipennis Fairmaire) (Coleoptera: Buprestidae) in North America: Results of multiple-choice field experiments. *Environ. Entomol*. 37(1): 230–241.

Baker, T. C. (2009). Use of pheromones in IPM. In: E. B. Radcliffe, W. D. Hutchison and R. E. Cancelado (Eds.) *Integrated Pest Management* (pp. 273–285). Cambridge, MA: Cambridge University Press.

Bakke, A. and Kvamme, T. (1981). Kairomone response in Thanasimus predators to pheromone components of Ips typographus. *J. Chem. Ecol*. 7(2): 305–312.

Ball, J. and Simmons, G. (1986). The influence of host condition on post 1st instar development of the bronze birch borer, Agrilus anxius (Coleoptera, Buprestidae). *Great Lakes Entomol*. 19: 73–76.

Bartelt, R. J., Cossé A. A., Zilkowski, B. W., and Fraser, I. 2007. Antennally active macrolide from the emerald ash borer Agrilus planipennis emitted predominantly by females. *J. Chem. Ecol*. 33(7): 1299-1302.

Bellamy, C. L. (2008). *World Catalogue and bibliography of jewel beetles* (Coleoptera: Buprestoidea). Sofia, Bulgaria: Pensoft.

Berryman, A. A., Dennis, B., Raffa, K. F., and Stenseth, N. C. (1985). Evolution of optimal group attack, with particular reference to bark beetles (Coleoptera, Scolytidae). *Ecology* 66(3): 898–903.

Bierl, B. A., Beroza, M. and Collier, C. W. (1970). Potent sex attractant of the gypsy moth: Its isolation, identification, and synthesis. *Science*. 170: 87–89.

Billings, R. F. and Cameron, R. S. (1984). Kairomonal responses of Coleoptera, Monochamus titillat*or* (Cerambycidae), Thanasimus dubius (Cleridae), and Temnochila virescens (Trogositidae), to behavioral chemicals of southern pine bark beetles (Coleoptera, Scolytidae). *Environ. Entomol*. 13(6): 1542–1548.

Burke, H. E. (1919). Biological notes on the flatheaded apple tree borer (Chrysobothris femoitata Fab.) and the pacific flatheaded apple tree borer (Chrysobothris malihorn). *J. Econ. Entomol*. 12: 326–330.

Butenandt, A., Beckmann, R., Stamm, D., and Hecker, E. (1959). Über den Sexuallockstoff des Seidenspinners Bombyx mori. Reindarstellung und Konstitution. Z. *Naturforsch*. 14b: 283–284.

Cappaert, D., McCullough, D. G., Poland, T. M., and Siegert, N. W. (2005). Emerald ash borer in North America: A research and regulatory challenge. *Am. Entomol.* 51(3): 152–165.

Careless, P. and Marshall, S. A. (2010). Working with the smoky-winged beetle bandit. *Bull. Entomol. Soc. Can.* 42(1): 25–29.

Carlson, R. W. and Knight, F. B. (1969). Biology, taxonomy, and evolution of four sympatric *Agrilis* beetles (Coleoptera: Buprestidae). *Contrib. Am. Entomol. Inst.* 4(3): 1–105.

Coleman, T. W. and Seybold, S. J. (2008). Previously unrecorded damage to oak, Quercus spp., in southern California by the goldspotted oak borer, Agrilus coxalis Waterhouse (Coleoptera: Buprestidae). *Pan-Pac. Entomol.* 84(4): 288–300.

Coleman, T. W. and Seybold, S. J. (2011). Collection history and comparison of the interactions of the goldspotted oak borer, Agrilus auroguttatus Schaeffer (Coleoptera: Buprestidae), with host oaks in southern California and southeastern Arizona, U.S.A. *Coleopt. Bull.* 65(2): 93–108.

Coleman, T. W., Grulke, N. E., Daly, M., Godinez, C., Schilling, S. L., Riggan, P. J., and Seybold, S. (2011). Coast live oak, Quercus agrifolia, susceptibility and response to goldspotted oak borer, Agrilus auroguttatus, injury in southern California. *Forest Ecol. Manag.* 261(11): 1852–1865.

Cossé, A. A., Bartelt, R. J., Zikowski, B. W., and Fraser, I. (2008). Identification and antennal electrophysiology of ash bark volatiles for the emerald ash borer. In: V. Mastro, D. Lance, R. Reardon and G. Parra (Eds.). *Proceedings of the Emerald Ash Borer and Asian Longhorned Beetle Research and Technology Development Meeting, Pittsburgh*, Pa., 23–24 October 2007. FHTET-2008-07. (pp. 81–82). Morgantown, WV: USDA Forest Service Forest Health Technology Enterprise Team.

Cote, W. A. and Allen, D. C. (1980). Biology of two-lined chestnut borer, Agrilus bilineatus, in Pennsylvania and New York. *Ann. Entomol. Soc. Am.* 73: 409-413.

Crook, D. J., Khrimian, A., Francese, J. A., Fraser, I., Poland, T. M., Sawyer, A. J., and Mastro, V. C. (2008). Development of a host-based semiochemical lure for trapping emerald ash borer Agrilus planipennis (Coleoptera: Buprestidae). *Environ. Entomol.* 37: 356–365.

Crook, D. J., Francese, J. A., Zylstra, K. E., Fraser, I., Sawyer, A. J., Bartels, D. W., Lance, D. R., and Mastro, V. C. (2009). Laboratory and field response of the emerald ash borer (Coleoptera: Buprestidae), to selected regions of the electromagnetic spectrum. *J. Econ. Entomol.* 102(6): 2160–2169.

Csóka, Gy. and Hirka, A. 2006. 2004 – Year of the gypsy moth in Hungary: Biotic damage in forests. In: Gy. Csóka, A. Hirka and A. Koltay (Eds.) *Proceedings of the IUFRO WP.* (pp. 271–275). Mátrafüred, Hungary. IUFRO.

Csóka Gy. and Kovács T. (1999). *Xilofág Rovarok – Xylophagous Insects*. Hungarian Forest Research Institute, Erdészeti Turományos Intézet. Budapest, Hungary: Agroinform Kiadó.

Davis, D. W. and Raghuvir, N. N. (1964). The biology of the rose stem girdler, Agrilus rubicola Communis, on raspberries in Utah (Coleoptera: Buprestidae). *Ann. Entomol. Soc. Am.* 57(2): 154–159.

De Groot, P., Grant, G. G., Poland, T. M., Scharbach, R., Buchan, L., Nott, R. W., Macdonald, L., and Pitt, D. (2008). Electrophysiological response and attraction of emerald ash borer to green leaf volatiles (GLVs) emitted by host foliage. *J. Chem. Ecol.* 34:1170–1179.

Domingue, M. J. (2011). Trapping of European buprestid species in Hungarian oak forests. In: D. Lance, and G. Parra (Eds.). *Proceedings of the Emerald Ash Borer and Asian Longhorned Beetle Research and Technology Development Meeting,* Wooster, OH., 5–6 October 2011. FHTET-(In press), Morgantown, WV: USDA Forest Service Forest Health Technology Enterprise Team.

Domingue, M. J., Csóka, Gy., Tóth, M., Vétek, G., Pénzes, B., Mastro, V., and Baker, T. C. (2011). Field observations of visual attraction of three European oak buprestid beetles toward conspecific and heterospecific models. *Entomol. Exp. Appl.* 140(2): 112–121.

Douglas, A. M. (1980). Our dying fauna. Perth, Australia: Imperial Publishing Company.

Dunn, J. P. and Potter, D. A (1988). Evidence for sexual attraction by the twolined chestnut borer, Agrilus bilineatus (Weber) (Coleoptera, Buprestidae) *Can. Entomol.* 120(11): 1037–1039.

Dunn, J. P. and Potter, D. A. (1991). Synergistic effects of oak volatiles with ethanol in the capture of saprophagous wood borers. *J. Entomol. Sci.* 26(4): 425–429.

Dunn, J. P., Kimmerer, T. W. and Nordin, G. L. (1986a). The role of host tree condition in attack of white oaks by the twolined chestnut borer, Agrilus bilineatus (Weber) (Coleoptera: Buprestidae). *Oecologia* 70(4):596–600.

Dunn, J. P., Kimmerer, T. W. and Nordin, G. L. (1986b). Attraction of the twolined chestnut borer, Agrilus bilineatus (Weber) (Coleoptera: Buprestidae), and associated borers to volatiles of stressed white oak. *Can. Entomol.* 118: 503–509.

Durrrer, H. and Villiger W. (1972). Schillerfarben von Euchroma gigantea (L.): (Coleoptera: Buprestidae): Elektronenmikroskopische untersuchung der elytra. *Int. J. Insect. Morphol.* 1(3): 233–240.

Evans, H. E. (1971). Observations on the nesting behavior of wasps of the tribe Cercerini. *J. Kansas. Entomol. Soc.* 44(4): 500–523.

Francese, J. A., Mastro, V. C., Oliver, J. B., Lance, D. R., Youssef. N., and Lavallee, S. G. (2005). Evaluation of colors for trapping Agrilus planipennis (Coleoptera: Buprestidae). *J. Entomol. Sci.* 40: 93–95.

Francese, J. A., Oliver, J. B., Fraser, I., Lance, D. R. Youssef, N., Sawyer, A. J., and Mastro, V. C. (2008). Influence of trap placement and design on capture of the emerald ash borer (Coleoptera: Buprestidae). *J. Econ. Entomol.* 101: 1831–1837.

Francese, J. A., Crook, D. J., Fraser, I., Lance, D. R., Sawyer, A. J., and Mastro, V. C. (2010a). Optimization of trap color for emerald ash borer (Coleoptera: Buprestidae). *J. Econ. Entomol.* 103: 1235–1241.

Francese, J. A., Fraser, I., Rietz, M. L., Crook, D. J., Lance, D. R., and Mastro, V. C. (2010b). Relation of prism trap color, size and canopy placement in determining capture of emerald ash borer (Coleoptera: Buprestidae). *Can. Entomol.* 142: 596–600.

Francese, J. A., Fraser, I., Lance, D. R., and Mastro, V. C. (2011). Efficacy of multifunnel traps for capturing emerald ash borer (Coleoptera: Buprestidae): Effect of color, glue, and other trap coatings. *J. Econ. Entomol.* 104(3): 901–908.

Grant, G. G., Ryall, K. L., Lyons, D. B., and Abou-Zaid, M. M. (2010). Differential response of male and female emerald ash borers (Col., Buprestidae) to (Z)-3-hexenol and manuka oil. *J. Appl. Entomol.* 134: 26–33.

Grant, G. G., Poland, T. M., Ciaramitaro, T., Lyons, D. B., and Jones, G. C. (2011). Comparison of male and female emerald ash borer (Coleoptera: Buprestidae) responses to phoebe oil and (Z)-3-hexenol lures in light green prism traps. *J. Econ. Entomol.* 104(1): 173–179.

Gwynn, D. T. and Rentz, D. C. F. (1983). Beetles on the bottle: Male buprestids mistake stubbies for females (Coleoptera). *J. Aust. Entomol. Soc.* 22: 79–80.

Haack, R. A. and Benjamin, D. M. (1982) The biology and ecology of the twolined chestnut borer, Agrilus bilineatus (Coleoptera: Buprestidae), on oaks, Quercus spp., in Wisconsin. *Can. Entomol.* 114: 385–396.

Haack R. A., Jendek E., Liu H., Marchant K. R., Petrice T. R., Poland T. M., and Ye, H. (2002). The emerald ash borer: A new exotic pest in North America. *Newsl. Mich. Entomol. Soc.* 47: 1–5.

Haack R. A., Petrice T. R. and Zablotny J. E. (2009). First report of the European oak borer, Agrilus sulcicollis (Coleoptera: Buprestidae), in the United States. *Great Lakes Entomol.* 42: 1–7.

Hansen J., Moulton J. K. and Klingeman ,W. E. (2010). Range expansion and adult flight activity of Agrilus subrobustus (Coleoptera: Buprestidae) in Tennessee. *Fla. Entomol.* 93(3): 444–445.

Hespenheide, H. A. and Bellamy, C. L. (2009). New species, taxonomic notes, and records for Agrilus Curtis (Coleoptera: Buprestidae) of Mexico and the United States. *Zootaxa* 2084: 50–68.

Hoebeke E. R. and Wheeler A. G. Jr. (2011). Agrilus subrobustus Saunders (Coleoptera: Buprestidae): New southeastern us records of an Asian immigrant on mimosa, Albizia julibrissin (Fabaceae) Proc. *Entomol. Soc. Wash.* 113(3): 315–324.

Jendek, E. (1994). Studies in the East Palearctic species of the genus Agrilus Dahl, 1823 (Coleoptera:Buprestidae). Part 1. *Entomol. Prob.* 25: 9 –25.

Jendek, E. and Grebennikov, V. V. (2009). Agrilus sulcicollis (Coleoptera: Buprestidae), a new alien species in North America. *Can. Entomol.* 141: 236–245.

Johnson, D. T. (1992). Rednecked cane borer galling affects blackberry growth and yield. *J. Am. Soc. Hortic. Sci.* 117(5): 780–783.

Johnson, D. T. and Mayes, R. L. (1989). Biology and control of the rednecked cane borer, Agrilus ruficollis (F.) (Coleoptera, Buprestidae), on blackberries in Arkansas. *J. Entomol. Sci.* 24(2): 204–208

Johnson, W. T., Lyon, H. H., Koehler, C. S., and Weidhass, J. A. (1991). *Insects That Feed on Trees and Shrubs*. (pp 270-271). Ithaca, NY: Cornell University Press.

Jones, E. A., Reed, D. D., Mroz, G. D., Liechty, H. O., and Cattelino, P. J. 1993. Climate stress as a precursor to forest decline - paper birch in northern Michigan, 1985-1990. *Can. J. For. Res.* 23(2): 229–233.

Kolk, A. and Starzyk, J. R. (1996) *Atlas Szkodliwych Owadow Lesnych*. Warsaw, Poland: Multico.

König, J. (1996). Situation und Ursachenanalyse der Eichenschäden In: A. Wulf and R. Kehr (Eds.) Sachsen. *Eichensterben in Deutschland: Situation, Ursachenforschung und Bewertung – Mitteilungen aus der Biologischen Bundesanstalt für Land- und Forstwirtschaft* (pp. 24–31). Berlin, Germany: Parey Buchverlag.

Kovacs, K. F., Haight, R. G., McCullough, D. G., Mercader, R. J., Siegert, N. W., and Liebhold, A. M. (2010). Cost of potential emerald ash borer damage in U.S. communities, 2009–2019, *Ecol. Econ*. 69(3): 569–578.

Kovacs, K. F., Mercader, R. J., Haight, R. G., Siegert, N. W., McCullough, D. G., and Liebhold, A. M. (2011).The influence of satellite populations of emerald ash borer on projected economic costs in U.S. communities, 2010–2020. *J. Environ. Manage*. 92(9): 2170–2181.

Lakatos, F. and Molnár, M. 2009. Mass Mortality of Beech (Fagus sylvatica L.) in South-West Hungary. *Acta Silv. Lign. Hung*. 5:75–82,

Lelito, J. P. (2009). The mating systems of the emerald ash borer and related buprestid beetles. Ph.D. dissertation. University Park, Pensylvania: The Pennsylvania State University.

Lelito, J. P., Fraser I., Mastro, V. C., Tumlinson, J. H., Böröczky, K., and Baker, T. C. (2007). Visually mediated 'paratrooper copulations' in the mating behavior of Agrilus planipennis (Coleoptera: Buprestidae), a highly destructive invasive pest of North American ash trees. *J. Insect Behav*. 20: 537–552.

Lelito, J. P., Fraser, I., Mastro, V. C., Tumlinson, J. H., and Baker, T. C. (2008). Novel visual-cue-based sticky traps for monitoring of emerald ash borers, Agrilus planipennis (Coleoptera: Buprestidae). *J. Appl. Entomol*. 132: 668–674.

Lelito, J. P, Böröczky, K., Jones, T. H., Fraser, I., Mastro, V. C., Tumlinson, J. H., and Baker, T. C. (2009). Behavioral evidence for a contact sex pheromone component of the emerald ash borer, Agrilus planipennis Fairmaire. *J. Chem. Ecol*. 35: 104–110.

Lelito, J. P., Domingue, M. J., Fraser, I., Mastro, V. C., Tumlinson, J. H., and Baker, T. C. (2011). Field investigations of the mating behaviors of Agrilus cyanescens and Agrilus subcinctus. *Can. Entomol*. 143(4): 370–37.

Lindgren, B. S. (1983). A multiple funnel trap for scolytid beetles (Coleoptera). *Can. Entomol*. 115: 299–302.

MacFarlane, D. and Meyer, S. P. (2005). Characteristics and distribution of potential ash tree hosts for emerald ash borer. *Forest Ecol. Manag*. 213:15–24.

Marshall, J. M., Storer, A. J., Fraser, I., and Mastro, V. C. (2010). Efficacy of trap and lure types for detection of Agrilus planipenn*is* (Col., Buprestidae) at low density. *J. Appl. Entomol*. 134(4): 296–302.

Marshall, J. M. Storer, A. J., Fraser, I., and Mastro, V. C. (2011). A predictive model for detection of Agrilus planipennis (Col., Buprestidae) larvae in girdled ash (Fraxinus spp.) *J. Appl. Entomol*. 135 (1-2): 91–97.

McCullough, D. G., Poland, T. M. and Cappaert, D. (2009). Emerald Ash Borer (Agrilus planipennis) attraction to ash trees stressed by girdling, herbicide or wounding. *Can. J. For. Res*. 39: 1331–1345.

McManus, M. and Csóka, Gy. (2007). History and impact of gypsy moth in North America and comparison to the recent outbreaks in Europe. *Acta Silv. Lign. Hung*. 3: 47–64.

Miller, R. O., Bloese, P. D., Hanover, J. W., and Haack, R. A. (1991). Paper birch and European white birch vary in growth and resistance to bronze birch borer. *J. Am. Soc. Hort. Sci*. 116(3): 580–584.

Moraal, L. G. and Hilszczanski, J. (2000). The oak buprestid beetle, Agrilus biguttatus (F.) (Coleoptera: Buprestidae), a recent factor in oak decline in Europe. *Journal of Pest Science* 73: 134–138.

Muirhead, J. R., Leung, B., Van Overdijk, C., Kelly, D. W., Nandakumar, K., Marchant, K. R., MacIsaac, H. J. (2006). Modelling local and long-distance dispersal of invasive emerald ash borer Agrilus planipennis (Coleoptera) in North America. *Divers. Distrib.* 12(1): 71-79.

Muskovits, J. and Hegyessy, G. (2002). Magyarország Díszbogarai – Jewel Beetles of Hungary. Nagykovácsi, Hungary: Grafon Kiadó.

Muzika, R. M., Liebhold, A. M. and Twery, M. J. (2000). Dynamics of twolined chestnut borer Agrilus bilineatus as influenced by defoliation and selection thinning. *Agric. For. Entomol. Entomology* 2:283–289.

Nielsen, D. G., Muilenburg, V. L. and Herms, D. A. (2011). Interspecific variation in resistance of Asian, European, and North American birches (Betula spp.) to bronze birch borer (Coleoptera: Buprestidae) *Environ. Entomol.* 40(3): 648–653.

Petrice, T. R. and Haack, R. A. (2006). Effects of cutting date, outdoor storage conditions, and splitting on survival of Agrilus planipennis (Coleoptera : Buprestidae) in firewood logs. *J. Econom. Entomol.* 99(3): 790–796.

Pitman, G. B., Vite, J. P., Kinzer, G. W., and Fentiman, A. F. (1968). Bark beetle attractants: trans-verbenol isolated from Dendroctonus. *Nature*. 218: 168–169.

Poland, T. M. and McCullough, D. G. (2006). Emerald ash borer: invasion of the urban forest and the threat to North America's ash resource. *J. Forestry* 104: 118–124.

Pureswaran, D. S. and Poland, T. M. (2009). The role of olfactory cues in short-range mate finding by the emerald ash borer, Agrilus planipennis Fairmaire (Coleoptera: Buprestidae) *J. Insect Behav.* 22(3): 205–216.

Raffa, K. F. and Berryman, A. A. (1983). The role of host plant-resistance in the colonization behavior and ecology of bark beetles (Coleoptera, Scolytidae). *Ecol. Monogr.* 53(1): 27–49.

Rodriguez-Saona, C., Poland, T. M., Miller, J. R., Stelinski, L. L., Grant, G. G., De Groot, P., Buchan, L., and Macdonald, L. (2006). Behavioral and electrophysiological responses of the emerald ash borer, Agrilus planipennis, to induced plant volatiles of Manchurian ash, Fraxinus mandshurica. *Chemoecology* 16: 75–86.

Roelofs, W. L. and Arn, H. (1968). Sex attractant of red-banded leaf roller moth. *Nature* 219: 513.

Roelofs, W. L., Comeau, A. and Selle, R. (1969). Sex pheromone of oriental fruit moth. *Nature* 224: 723.

Rudinsky, J. A. (1966). Scolytid beetles associated with Douglas fir: Response to terpenes. *Science*. 152: 218–219.

Ryall, K. L., Fidgen, J. G. and Turgeo, J. J. (2011). Detectability of the emerald ash borer (Coleoptera: Buprestidae) in asymptomatic urban trees by using branch sample. *Environ. Entomol.* 40(3): 679-688.

Sargent, C., Raupp, M., Bean, D., and Sawyer, A. J. (2010). Dispersal of emerald ash borer within an intensively managed quarantine zone. *Arbor. Urb. For.* 36(4): 160–163.

Schlag, M. G. (1992). Oak decline in Europe and its causes as seen from a phytopathological point of view. *Austrian. J. For. Sci.* 111: 243–266.

Schopf, A. (1992). Rinden- und holzbrütende Schädlinge an erkrankten Eichen in Österreich. *Österr. Forstzeitung*. 1: 33–35.

Schultz, T. D. and Rankin, M. A. (1985). The ultrastructure of epicuticular interference reflectors of tiger beetles (Cicindela) *J. Exp. Biol.* 117: 88–110.

Seago, A. E. Brady, P., Vigneron, J. P., and Schultz, T. D. (2009). Gold bugs and beyond: A review of iridescence and structural colour mechanisms in beetles (Coleoptera). *J. Roy. Soc. Interface*. 6: S165–S184.

Silk, P. J., Ryall, K., Lyons, D. B., Sweeney, J. D., and Wu, J. P. (2009). A contact sex pheromone component of the emerald ash borer Agrilus planipennis Fairmaire (Coleoptera: Buprestidae). *Naturwissenschaften* 96(5): 601–608.

Silk, P. J., Ryall, K., Mayo, P., Lemay, M. A., Grant, G., Crook, D., Cossé, A., Fraser, I., Sweeney, J. D., Lyons, D. B., Pitt, D., Scarr, T., and Magee, D. (2011). Evidence for a volatile pheromone in Agrilus planipennis Fairmaire (Coleoptera: Buprestidae) that increases attraction to a host foliar volatile. *Environ. Entomol*. 40(4): 904–916.

Silverstein, R. M., Rodin, J. O. and Wood, D. L. (1966). Sex attractants in frass produced by male Ips confusus in ponderosa pine. *Science* 154: 509–510.

Silverstein, R. M., Brownlee, R. G., Bellas, T. E., Wood, D. L., and Browne, L. E. (1968). Brevicomin: Principal sex attractant in the frass of the female western pine beetle. *Science*. 159: 889–891.

Smitley, D., T. Davis and Rebek, E. (2008). Progression of ash canopy thinning and dieback outward from the initial infestation of emerald ash borer (Coleoptera: Buprestidae) in southeastern Michigan. *J. Econ. Entomol*. 101: 1643–1650.

Tumlinson, J. H., Hardee, D. D., Gueldner, R. C., Thompson, A. C. Hedin, P. A., and Minyard, J. P. (1969). Sex pheromones produced by male boll weevil - isolation, identification, and synthesis. *Science* 166: 1010.

Turnbow, R. H. and Franklin, R. T. (1983). Bionomics of Brachys tessellatus (Coleoptera, Buprestidae) in coastal-plain scrub oak communities. *Ann. Entomol. Soc. Am*. 74(4): 351-358.

Vansteenkiste, D., Tirry, L., Van Acker, J., and Stevens, M. (2005). Predispositions and symptoms of Agrilus borer attack in declining oak trees. *Ann. For. Sci*. 61: 815–823.

Wang, X. Y., Yang, Z. Q., Gould, J. R., Zhang, Y. N., Liu, G. J., and Liu, E. S. (2010). The biology and ecology of the emerald ash borer, Agrilus planipennis, in China. *J. Insect Sci*. 10: 128.

Wargo, P. M. (1977). Armillariella mellea and Agrilus bilineatus and mortality of defoliated oak trees. *Forest Sci*. 23(4): 485–492.

Wargo, P. M. and Montgomery, M. E. (1983). Colonization by Armillaria mellea and Agrilus bilineatus of oaks injected with ethanol. *Forest Sci*., 29(4): 848–857.

Westcott, C. (1973). *The gardner's bug book*. 4th ed. (689 pp.) Garden City, NY: Doubleday and Co.

Wood, D. L., Browne, L. E., Bedard, W. D., Tilden, P. E., Silverstein, R. M., and Rodin, J. O. (1968). Response of Ips confusus to synthetic sex pheromone in nature. *Science*. 159: 1373-1374

Yu, C. 1992. Agrilus marcopoli Obenberger (Coleoptera: Buprestidae). In: G. Xiao (Ed.) *Forest insects of China*, 2nd Ed. (pp. 400–401). Beijing, China: China Forestry Publishing House.

In: Invasive Species
Editors: Joaquín J. Blanco and Adrian T. Fernandes

ISBN: 978-1-61942-761-7

Chapter 4

THE INVASIVE BEAVER *CASTOR CANADENSIS* IN THE TIERRA DEL FUEGO ARCHIPELAGO: A MITOCHONDRIAL DNA AND SPATIAL GENETIC STRUCTURE ANALYSIS FOR CONTROLLING POPULATION EXPANSION

Mariana Fasanella and Marta Susana Lizarralde*
Centro Regional de Estudios Genómicos, UNLP.
Buenos Aires, Argentina

ABSTRACT

The Tierra del Fuego Archipelago (TDFA) contains numerous invasive species, of which *Castor canadensis* is the most abundant and important invasion. Beaver are responsible for the most drastic landscape alteration in Tierra del Fuego since the last glacial age, affecting not only the hydrology and composition of the southern beech forest, but more importantly allowing other exotic species to invade the ecosystem. From 25 pairs intentionally released in 1946, beavers have increased their numbers to a current population size of aprox.100,000 individuals. Here, we present the genetic variability and population structure of 222 mitochondrial DNA beaver samples. We detected 7 D-loop haplotypes, 3 of them resulted the most abundant and distributed all along the archipelago. To analyze the data, TDFA population was subdivided into five subpopulations and a slight structure was found (ϕst =0.169). With this data, we also identify management units (MUs) of the invasive beaver in the TDFA. Although, sea water and mountains could be considered barriers for beaver's dispersion, we found no barriers to gene flow within the Isla Grande, which may be due to the complex stream watershed network that allows beavers to easily disperse. These results indicate that the Isla Grande is a single management unit (MU) and every small island in the archipelago is a separate MU. It is important to consider that successful eradication programs of mammals were always conducted in small islands while invasive populations on larger islands or those that display no distinct structure are more problematic, so the eradication

* mlizarralde@fibertel.com.ar.

of *Castor canadensis* in the TDFA would be logistically very difficult or perhaps impossible. Therefore, we propose to (1) control the expansion from IG population to others areas through trapping beavers and (2) eradicate beaver populations of the adjacent small islands (small area comparative to Isla Grande) and especially from the continent. If the species is not removed early in the mainland, beavers will begin to invade the Patagonian forests very quickly and its future eradication in these continental areas will be almost impossible. To carry out the control of the species in Tierra del Fuego, is necessary to promote beaver extraction through the commercial trapping in a way to control this species in the short term. To achieve this important point, the classic control scheme must be change it stimulating new business development by incorporating new materials and products to market and also coordinate the participation of public and private actions to exploit species under control. In this work, we state three options for economic use of the beaver: 1) the government could pay a price for killing beavers, 2) to sell/export the beaver fur and 3) to sell beaver´s meat for human consumption. With these three options of beaver exploitation, we believe that the Isla Grande control may be feasible and logistically possible.

Introduction

Historical Aspects: Introduction and Colonization of Tierra Del Fuego Archipelago

Species introductions is one of the most controversial topics in Ecology, since Elton's (1958) pioneering work, with implication in biological conservation, mostly due to the biodiversity loss it enthralls (Sala *et al.* 2000; Vázquez 2002), and the great economic losses it generates (Pimentel *et al.* 2000; Curtis and Jensen 2004). Exotic invasive species are currently the second cause of threats to and extinction of native species, only surpassed by habitat loss (UICN). Among the many exotic species introduced to new habitats, an estimated 10% have the ability to increase their populations, becoming successful invaders (Kolar and Lodge 2001). The North American beaver (*Castor canadensis*) is one of them.

In November 1946, twenty-five mating pairs of *C. canadensis* (Kuhl 1820) were introduced from Canada and released into Río Claro, located on the Isla Grande de Tierra del Fuego (TDF). Such an introduction was made with the purpose of creating a fur industry, and was backed at the time by Argentina's Marine Ministry. The absence of most natural predators, no laws regulating hunting, and the rich abundance of forage and habitats, led to a rapid increase in the North American beaver populations (Lizarralde 1993). Besides, it is worth mentioning that island ecosystems (such as Tierra del Fuego Archipelago, TDFA) have few plants, herbivores, carnivores and decomposers playing active parts in ecological processes, and therefore are more vulnerable to species invasions (Lever 1994). The geographical progression of the species was noted on numerous sites throughout the years. For instance, in 1964, beavers were present on the Chilean side of Lago Fagnano (western extreme), and by the mid-1960s, beavers had crossed the Beagle Channel, invading Isla Navarino (Anderson *et al. 2011*). The colonization of the section of Parque Nacional Tierra del Fuego located south of Lago Fagnano, probably took place during the 1970s, given the absence of colonies or impact signs in aerial photographs of the area dating from 1970 (Lizarralde 1989, 1993). In 1979, beavers were found in San Sebastián and later on, by 1986 they were found in the area of China Creek, Puesto Calafate and Río Marazzi (Chile). The original population progressively invaded the neighboring islands of Navarino, Hoste, Picton, Nueva and Lennox (Anderson *et al.* 2009), reaching continental areas by the mid-1990s, most notably on the Brunswick Peninsula in Chile (Wallem *et al.* 2007). To date, the presence of beavers has not been reported for the Wollaston islands (Parque Nacional Cabo de Hornos, Chile), nor for the western sections of the Archipelago (Parque Nacional D´Agostini, Chile) (Anderson *et al.* 2006; Moorman *et al.* 2006). Even though it has been proposed that the optimum habitat for the species on the Isla Grande is the subantarctic *Nothofagus* forest (Lizarralde 1993; Lizarralde *et al.* 1996, 2004, 2008; Martínez Pastur *et al.* 2006; Wallem *et al.* 2007; Anderson *et al.* 2009), beavers have been able to colonize open spaces such as the Fuegian ecotone (located in the center of Isla Grande), and the Magellanic steppe (northern Isla Grande). This is in agreement with studies performed in the Northern hemisphere, where the North American beaver showed a preference for pasture-type environments, with no woody species (Barnes and Malik 1997; Susuki and McComb 1998).

Skewes *et al.* (1999) estimated the population advance rate of the North American beaver, considering the year of introduction of beavers into TDF and calculating the lineal distance from the original source point (Río Claro basin) to those locations where populations had been reported. For instance, the population advance rate to the rural outpost in Fagnano was 3.8 km yr^{-1}, while in the northern part of the island, beavers presented an advance rate of 6.3 km yr^{-1}, a much higher rate compared to other environments. Even if these population advance rates are just approximate, it can be noted that the rate has not been uniform among different geographic areas, with higher rates in northern (6.3 km yr^{-1}) and southern (5.7 km yr^{-1}) areas. These geographical variations in advance rate can be explained as a function of the main niche dimensions of the species, such as hydric (offering refuge and means of transport) and forest richness (food source). The higher rate in northern TDF, where the conditions for these two parameters would be most unfavorable, due to a scarcity of forests and its steppe character, is at first hand contradictory. However, the explanation could lie on the relatively high abundance of water streams, totaling 2,042 km, in contrast to 1,711 km in the central zone, and 3,630 km in the southwestern forested zone. On the other hand, the rapid advance in the northern zone could be attributed to the scarcity of food for foraging, and the difficulties in the construction of beaver lodges imposed by the landscape (*e.g.* this region only presents small shrubs); what ultimately may have forced beavers to look for better environments at a faster displacement rate. This situation would not necessarily take place in the southern zone where the North American beaver has abundant trees for foraging. The low population density in the northern zone, even if it presents plenty of water streams, indicates that the presence of water streams by itself is not enough for the establishment of numerically-important populations of the North American beaver (Skewes *et al.* 1999). The absence of forests is a limiting factor of its advancement (although not absolutely). Briones *et al.* (2001) established that the association between terrain slope, marginal vegetation, stream width and forest community composition in the riparian zone is important for the establishment of beaver populations. In Sweden, the rate and dispersal pattern of the European beaver is mainly associated with hydrographic basins (Hartman 1995), while in TDF, the absence of streams is no obstacle to its dispersal (Skewes *et al.* 1999).

The North American beaver seems to have attained its northern distribution limit in TDF; with the scarcity of water stream-associated plant biomass being the main limiting factor. The altitude distribution limit is determined by the availability of vegetation, especially trees; on the other hand, the North American beaver reaches down to river mouths, with beaver lodges and dams found only 50 m from the ocean. Its presence in brackish water has been corroborated and explains the colonization of other islands within the Archipelago (Ramírez Silva 2006). It is probable that those individuals that established the expansive south-west route, gave origin to the populations currently present at Isla Dawson. The highest risk of advancement of the invasive beaver population from this island towards the continent, has been predicted on the basis of the short distance separating both locations, and the high population density at Isla Dawson. In support of the latter, sporadic but continuous sightings of individuals on Lago Parrillar, located in Brunswick Peninsula (continental Chile), have taken place since 1994. The absence of predators in the zone might induce behavioral changes (such as an increment in foraging time), which might in turn increase the beavers' home range, leading to the exploitation of larger areas of forest by a single colony (Wallem *et al.* 2007). However, the presence of a great predator such as the puma (*Felis concolor*) in continental areas, might prevent its advancement and lower its chances of displacement for foraging and lodge construction.

Even though the provincial government of Tierra del Fuego started regulating commercial hunting of beavers in 1983, and diverse efforts were made to control populations, the species rapidly expanded its range throughout the TDFA nonetheless, currently inhabiting 98% of streams at Isla Grande, with an estimated abundance of 100,000 individuals (Lizarralde *et al.* 2008). Regardless of the lack of precise information on recent years colonization, it can be mentioned that the invasive dynamics of the North American beaver currently include the reoccupation of abandoned sites (due to dam overflows), the colonization of elevated sections of streams on mountain sides (not used during the initial stages of colonization), and the dispersal towards the steppe.

The invasion of the North American beaver and its impacts throughout the whole Archipelago has generated interests on the part of Argentinean and Chilean government agencies, conservation organizations and scientific institutions, agreeing on the need for the implementation and evaluation of control measures and its possible eradication from southern South America. In this scenario, designing an effective management plan in order to prevent the invasion of new sites, especially in the continent, becomes paramount.

Biological Aspects of *Castor canadensis*

The North American beaver is the second largest rodent, after the capybara (*Hydrochoerus hydrochaeris*); it is taxonomically placed in the Castoridae family, within the order Rodentia. There are currently two living species with allopatric distribution: *C. canadensis* in North America and, *C. fiber* in Europe and Asia. *C. canadensis* ranges in its original distribution from Alaska to the Gulf of Mexico (Müller-Schwarze and Sun 2003), although it has been introduced to other regions such as the Scandinavian Peninsula, Finland and Argentina (TDF) (Figure 1). As a consequence of the introduction of *C. canadensis*, *C. fiber* is currently displaced in certain areas of Europe and Asia (Durka *et al.* 2005).

Beavers are characterized by the construction of dams in association with flowing water and the formation of "beaver marshes" and "burrows" where they inhabit and store food. Due to the structural changes that it makes on the ecosystem, it is considered a "keystone species".

Specimens from TDF range in weight between 14 and 30 kg, with an average of 17.88 ± 7.51 kg. Total length is 120 cm, with a mean length of 101.85 ± 7.48 cm. The most remarkable characteristic of the species is the flattened, paddle-like tail, which may reach a length of 23-33 cm and a width of 11-18 cm (Lizarralde and Escobar 1999).

Beavers do not present sexual dimorphism, with internal sexual organs; therefore, the easiest way of sexing organisms is by palpation (for males), and during lactation (for females). In the Fuegian Archipelago, the reproductive period starts in June and extends to September, with a gestation period lasting 90-100 days. Kits (young-of-the-year) show up in early spring; the mean number of kits in each cohort is 3.37 animals, ranging between 3-7 kits. There are no records of beavers having more than a single litter per year (Lizarralde and Escobar 1999). Beavers have a social organization based on family groups (termed a "colony" here) that inhabit a single marsh or a succession of marshes along a stream and use a common feeding ground (termed "food catch"). The number of animals per colony is about 5 (range 4-6) (Lizarralde 1993, Lizarralde *et al.* 1996). Beavers are monogamous and a colony is comprised of a parental couple, the yearlings younger than 2 years, and the young-of-the-year. The number of individuals and composition of the colony can vary according to habitat quality, with some bigger colonies in forested areas and smaller ones in the steppe. Lizarralde (1993) and Briones *et al.* (2001) reported a colony density in forested areas ranging between 0.7 - 1 colonies km^{-1} for Isla Grande, while in Isla Navarino the density was of 1.1 colonies km^{-1} (Skewes *et al.* 2006). The diet is exclusively vegetarian and includes bark, leaves and twigs of woody species; beavers are generalist herbivores that consume at least 80 woody species and 149 herbaceous plants in their native range (Collen and Gibson 2001). Plants serve mainly three functions: 1) a direct source of food, 2) food reserves for the winter, and 3) construction materials for beaver lodges and dams (Busher 1996). In Tierra del Fuego, the summer diet of beavers is mainly composed of herbaceous species, while in winter they turn to woody species (Castillo 2006).

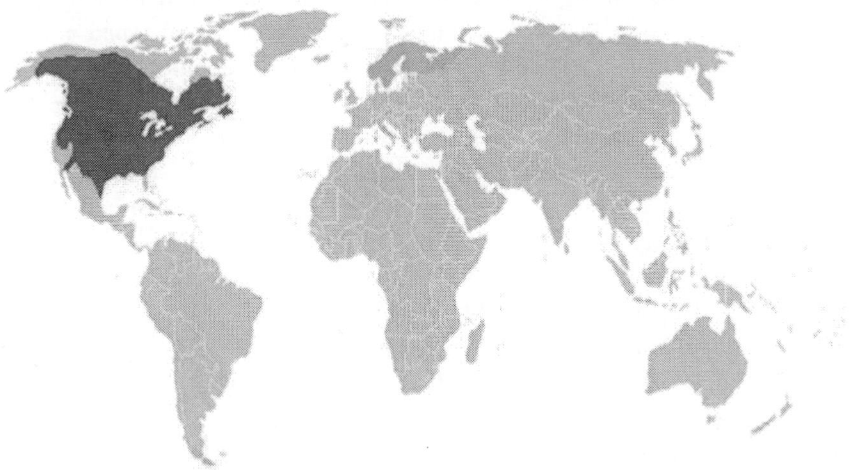

Source: Wikipedia.

Figure 1. *Castor canadensis* distribution. In dark green: native distribution. In light green: exotic population.

The ñire (*Nothofagus antarctica*), lenga (*N. pumilio*), guindo (*N. betuloides*) and shrubs of the genera *Pernettya*, *Berberis*, *Chiliotrichum*, and *Juncus* constitute the main plant species on which beavers forage in Tierra del Fuego. Tree-cutting activities in TDF are more intense during the fall, when the material is chipped and stored underwater, in the proximity of beaver lodges, as a food reserve for the winter (i.e., the so-called "food catches") (Lizarralde and Escobar 1999).

Beavers' teeth are characterized by long, sharp and strong incisives which enable them to chop down trees > 1 m in diameter. Additionally, the lips can be sealed behind the incisives, allowing them to chip wood underwater (Massoia and Chébez 1993; Müller-Schwarze and Sun 2003).

Effects of Alteration on the Fuegian Ecosystem

The ability to build dams and expand the marshy area increases the habitat and food availability for beavers, offering protection from predators such as wolves (*Canis lupus*) in the northern Hemisphere, and potentially foxes (*Pseudalopex culpaeus*) in the southern Hemisphere; the invasive beaver has not been found in the diet of foxes (Jaksic *et al.* 1983), but scars have been found in specimens from Isla Grande (Andrade 2005). The dams built by beavers modify the annual discharge regime of a river, they reduce current velocity contributing to the formation of a series of graded "steps"and they expand the area of flooded soils and increase sediment and organic matter retention (Naiman *et al.* 1988). The introduction of the North American beaver has undoubtedly had several ecosystem effects (some positive, some negative) in Tierra del Fuego (Lizarralde 1993, Iriarte 2000, APN 2006).

Some of the most important negative effects are: 1) the destruction of the riparian forest, which produces soil destabilization and marked erosion effects on the rest of the forest; 2) as a consequence of the opening of new clearings, the light regime gets disrupted with high solar radiation, open terrain gets exposed to the erosive action of wind and rain, altogether modifying the microclimate of unexposed surfaces; 3) modifications of habitat structure and aquatic biota; 4) expansion of marsh areas, due to the changes in the drainage pattern and the height of aquifers; 5) changes in temperature, nutrients and water fluxes, negatively affecting native fish species; and 6) accumulation of sediment and organic matter, with the ultimate modification in nutrient cycling within water basins and adjacent areas (Lizarralde *et al.* 1996, 2004). Without any controls, these alterations could remain as part of the landscape for decades or centuries, maybe even indefinitely.

Plant litter resulting from chipping down trees, together with erosion of unforested areas result in an accumulation of organic and inorganic products that acidify the water and bottom of beaver dams and its surrounding area (Lizarralde *et al.* 1996). Moreover, the natural dynamics of the forest are seriously impacted, given that the recovery rate in newly flooded areas is very slow. Other negative impacts are evidenced by a loss of trees and modification of soil properties, either through a direct (trees as a food source, or trees fallen to build dams), or indirect actions (flooding). *Nothofagus* forests are the most affected ecosystem, especially the highly productive forests of *N. pumilio*, since the lenga is the main food source for the North American beaver in TDF. In all Patagonia, forests of *N. pumilio* and *N. betuloides* can adapt to large-scale alterations such as fires, wind- or landslide-induced fall of trees, but they seem to be extremely vulnerable to beaver effects (Silva and Saavedra 2008).

Another common disturbance is observed in roads, bridges, hedges and trenches, which are regularly weakened, interrupted or blocked by the action of the North American beaver. The dynamics generated by the use, exhaustion and abandonment of those sites occupied by beavers is associated with the later colonization by exotic plants. On the other hand, the fauna associated with these altered environments can be directly or indirectly impacted, as is the case of the torrent duck (*Mergannetta armata*) which has seen its range displaced to elevated portions of rivers and water streams (Massoia and Chébez 1993).

Among the positive impacts, the generation of useful environments for some bird species, especially within the order Anseriformes, is worth mentioning. These make use of the exposed parts of beaver lodges for nesting, and they profit from the clearing of parts of the dense forest which favor the growth of plantings and provide nesting areas for migrating birds. Wright *et al.* (2002) found that habitat modifications produced by the North American beaver, increased the number of herbaceous plant species growing on stream banks by 33%. Lastly, it was estimated that in a little more than half a century, the North American beaver has modified 25,000 ha of Fuegian forests, or roughly the equivalent forest biomass modified for agriculture practices (Lizarralde 1993).

Population Genetic Structure

Knowing the genetic structure of a certain species, makes it possible to identify the evolutionary unit and the Management Unit (MU) for it (Moritz 1994). "Management unit" is a category that makes reference to a group of individuals among which the degree of ecological and genetic connectivity is low enough to justify a separate monitoring and management for each group (subpopulation) separately (Palsbøll *et al.* 2006). Management units are useful to recognize demographically-distinct populations which need to be managed in order to ensure the viability of an Evolutionarily Significant Unit (ESU), which are bigger. In island populations, one of the biggest risks of failure of a management program (control - eradication) is the ability of species to recolonize from adjacent islands or even from the continent. Interconnected or geographically close groups of islands enabling migrations of specimens have been termed "Eradication Units" (EU) (Robertson and Gemmell 2004). These EU can be defined as genetically-isolated units with groups (or clusters) of populations which have to be eradicated at the same time in order to maximize the success of such strategy in the long run. Identifying EU is no easy task, given that migration patterns depend upon a number of biological, geographical and human factors. Therefore, the analysis of the population structure of a species between island clusters and interpreting this structure in terms of gene flow, is the adequate work frame for the identification of EU. Determining the inter-island migrating ability of an invasive species and delimiting EU makes eradication efforts more viable in the long run, since, in that way, a recolonization from neighboring islands is prevented.

Invasive species eradication is an important tool in ecological conservation and restoration. It provides the means to alleviate and/or remove the adverse effects that invasive exotic species typically generate in their new ecosystems. For an eradication program to be successful, the EU (i.e. the target population) must be clearly defined. Populations in small islands or those that are "isolated", are intrinsically well defined as EU (e.g. the eradication of mammals from coastal islands in New Zealand; Towns and Broome 2003), while populations in bigger islands, or those that do not present a structure, are more problematic when it comes to define EU and therefore, for the eradication of species. Even though large scale eradications are possible (Taylor *et al.* 2000; Towns and Broome 2003), they are logistically difficult to carry out and very cost-expensive.

A successful eradication strategy requires considerable planning beyond the mere identification of population units that are big enough and have low risk of recolonization. For example, trying to eradicate a fraction of the population or a "sink" population within an unidentified source-sink population dynamic, would inevitably lead to a rapid recolonization and therefore, an economic loss. It is worth mentioning that just a few migrants per cohort have the capacity to reduce the genetic differences between populations. Because of this, the successful eradication of a species requires the elimination of all individuals, or at least preventing the flux of migrants between populations. In that sense, natural dispersal routes are very important when defining eradication strategies. Therefore, the objectives of this chapter are:

1) to quantify levels of intra- and interpopulation genetic variation of *C. canadensis* in TDFA,
2) to characterize the population and genetic structure of the invasive species in TDFA, and
3) to propose potential management or eradication strategies for TDFA.

Materials and Methods

Study Area and Sampling

The study area comprised the whole Tierra del Fuego Archipelago (TDFA). Fresh-or alcohol-fixed tissue samples from 255 beavers were collected by hunters or donated by different Argentinean and Chilean organizations (see Acknowledgements section). Samples were georeferenced and deposited in the Sample Bank of Laboratorio de Ecología Molecular del Centro Regional de Estudios Genómicos (CREG), Universidad Nacional de La Plata, UNLP (Argentina).

DNA Extraction

DNA was extracted from fresh tissue samples (muscle, liver, spleen), and dry tail samples. For most tissues, DNA was extracted following protocols by Sambrook *et al.* (1989), with minor modifications in incubation times and concentrations, according to the type of tissue. For those samples of fresher tissue, we followed the protocol by Aljanabi and Martínez (1997). All DNA samples were deposited in the collection of Laboratorio de Ecología Molecular, CREG, UNLP and are preserved in eppendorf test tubes at -20°C for later genetic analyses.

Mitochondrial DNA Amplification and Sequencing

A D-loop fragment (500 bp) was amplified using the universal primers: Thr-L15926 (5′ CAATTCCCCGGTCTTGTAAACC-3′), located in proximity to the proline tRNA gene, and DL-H16340 (5′CCTGAAGTAGGAACCAGATG-3′), after Vilà *et al.* (1999). PCR was performed following the protocol in Fasanella *et al.* (2010). The amplification was done on a total volume of 25 μl with the following components: 25-100 ng DNA, 1x Buffer Taq Polymerase, 1.5 mM Cl_2Mg, 200 μM of each dNTP′s, 25 mM of primer and 1.25 U Taq Polymerase (PB-L^{TM}, INVITROGEN^{TM}). PCR was performed on an automated cycler THERMO HYBAID MBS o.2S and consisted of denaturation at 94°C for 30 sec, annealing at 55-57°C for 30 sec and extension at 72-74°C for 1-5 min, the cycle was repeated 40 times. For each individual, two PCR of 25 μl each were performed. Double-stranded PCR products were purified by precipitation and directly sequenced in both directions using the same primers for amplifications. Sequencing was conducted at MACROGEN Inc. (Korea). Sequences for both strands were determined and chromatograms were edited and aligned using the software program BIOEDIT 7.0 (Hall 1999).

Data Analysis

Genetic diversity of populations was studied estimating parameters of haplotype diversity (h) and nucleotide diversity (π) with the program DNAsp 4.10 (Rozas *et al.* 2003). Haplotype diversity is the probability for two DNAmt sequences randomly taken from a population to be different; while nucleotide diversity is the number of differences in nucleotides between two pairs of sequences taken randomly. An haplotype network was done following the "median-joining" method using the program NETWORK 4.6.0.0. (Fluxus Technology Ltd. 1999-2011). Haplotype networks (defined as plots connected by circles) are more appropriate to show relationships within a species.

Hierarchical analyses of molecular variance (AMOVA) were performed with the program GenAIEx 6.0 (Peakall and Smouse 2006) in order to study different levels of population structure. Such analysis considers genetic distances between haplotypes and their frequencies; groups of populations have to be defined *a priori* in order to evaluate if the genetic structure proposed by the operator is significant. AMOVA decomposes variance in: a) differences in the composition of haplotypes between individuals from different populations (*i.e.* variance within a population); b) differences in the composition of haplotypes of individuals from different populations (*i.e.* variance between populations); and c) differences in the composition of haplotypes between groups of populations (*i.e.* variance between regions). For this analysis, the population of *C. canadensis* (Argentina-Chile) was divided into 5 subpopulations defined in Figure 2.

In order to determine if there is any relationship between genetic distance and geographic distance, a Spatial Autocorrelation Analysis was performed for those samples from Parque Nacional Tierra del Fuego (PNTDF). PNTDF is located within the distribution range of subpopulation B, and is the only zone which has an abundant number of geographically-close samples (n = 72).

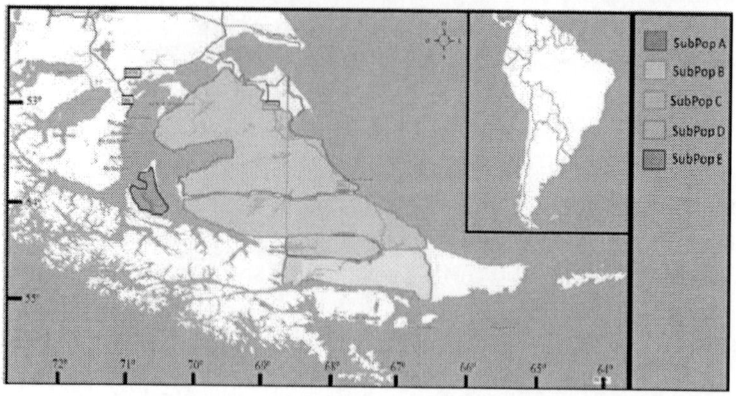

Figure 2. Location of the 5 subpopulations (A-E) in the Archipelago of Tierra del Fuego.

Correlation diagrams show the correlation index (r) which provide as proxy for genetic similarity between pairs of individuals whose spatial separation falls within a distance range. This analysis was performed with the program GenAIEx 6.0 (Peakall and Smouse 2006). Since there is no way of inferring genetic structure *a priori*, 9 distance classes were considered so that each of them would have a similar number of associated samples (*i.e.* 0-800, 801-1300, 1301-2000, 2001-3000, 3001-4000, 4001-5000, 5001-6000, 6001-7000, and 7001-10500 m). Significance tests for the "r" statistic were calculated for each distance class, defining the upper and lower limits for the confidence interval; this statistical procedure is described in Peakall and Smouse (2006). Positive values of "r" ($r > 0$) indicate a positive autocorrelation, meaning that those individuals genetically closer are also closer geographically. On the other hand, if $r < 0$, then the spatial autocorrelation is negative and therefore genetically similar individuals are further appart geographically.

In order to evaluate the relationship between haplotypes of *C. canadensis* from populations introduced to TDFA (Argentina-Chile) and northern Hemisphere populations, an haplotype network was done using Northern Hemisphere samples and D-loop haplotypes deposited in GenBank (Ducroz *et al.* 2005), with the software NETWORK 4.6.0.0. (Fluxus Technology Ltd., 1999-2011). For the northern Hemisphere, 15 samples from Alaska and Minnesota (USA) (see Acknowledgements section) and three haplotypes of *C. canadensis* deposited in GenBank (Ca1, Ca2 and Ca3; Ducroz *et al.* 2005) were used.

Results

A ~500 bp D-loop fragment was amplified and sequenced successfully in the 255 beaver samples considered. Haplotype (h) and nucleotide (π) diversity was 0.684 and 0.00444, respectively, indicating a high haplotype diversity and relatively moderate nucleotide diversity. The higher diversity haplotype (h) at the subpopulation level was found in subpopulation B ($h = 0.765$), the only subpopulation which presents the 7 haplotypes. On the other hand, values for nucleotide diversity oscillated between 0.00391 and 0.00503, which are very similar values. This index is related to the number of nucleotide differences between haplotypes (1.87 in average). For this analysis, the Archipelago was divided into five subpopulations according to the degree of advancement of beaver invasion (Anderson *et al.*

2009), the type of environment (steppe, ecotone or forest), the density of colonies km^{-1} and the type of hydrographic basin (Lizarralde 1993) (Figure 2). The following units or subpopulations were determined accordingly:

1. Subpop A (n = 67) representative of the "buffer" area surrounding the original introduction site (Río Claro), this is the founding population;
2. Subpop B (n = 87) representative of the forest and wetlands area with higher individual density (4.72 - 5.85 colonies km^{-1}; Lizarralde 1993);
3. Subpop C (n = 35) representative of an ecotone with intermediate densities (~ 2 colonies km^{-1}; Lizarralde 1993);
4. Subpop D (n = 45) representative of the steppe with wide rivers of low slope, where density is lower than in the other areas (0.2 colonies km^{-1}; Lizarralde 1993); and finally
5. Subpop E (n = 21) from Isla Dawson (Chile).

Seven DNAmt haplotypes (Haplotype A – F and Haplotype H, deposited in GenBank) were identified; they differed in six polymorphic sites in positions 39, 114, 128, 201, 212 and 236 (all transitions, Table 1). Haplotypes B (~ 40%), D (~ 20%) and H (~ 34%) were the most abundant ones, the rest of the haplotypes representing less than 2.5% (Table 1).

All subpopulations shared haplotypes B, D and H; two subpopulations shared four haplotypes (B, D, F and H), while only Subpop B (in Parque Nacional Tierra del Fuego) presented all of the haplotypes detected in this study.

Table 1. Polymorphic sites at the control region of mitochondrial DNA for *Castor canadensis*. GenBank accession numbers are shown

Nucleotide Position							
	39	114	128	201	212	236	
							GenBank Accession Number
HAPLOTYPE A	A	C	C	G	T	G	AY787822
HAPLOTYPE B						A	AY787823
HAPLOTYPE C	G						AY787824
HAPLOTYPE D			T	A		A	AY787835
HAPLOTYPE E			T	A			AY787826
HAPLOTYPE F		T		A	C		AY787827
HAPLOTYPE H		T		A	C	A	EU476079

The haplotype network (Figure 3) shows that three haplotypes were exclusive to Subpop B, while the rest of the haplotypes were shared by five subpopulations, except for haplotype F which was shared by Subpop A and B. Figure 3 also evidences that two hypothetic haplotypes would be the ancestral haplotypes, since they connect the rest of the haplotypes.

AMOVA evidenced a 6% difference in population structure between subpopulations, while within each subpopulation, diversity was 94% (Figure 4). The difference between the five subpopulations was $\phi st = 0.058$ ($p = 0.002$) indicating that the population is lightly structured. Pairwise, comparisons indicate a smaller gene flow between subpopulations B and C ($\phi st = 0.100$; $p = 0.001$), followed by B and E ($\phi st = 0.091$; $p = 0.007$), which are the subpopulations located further apart and separated by the Strait of Magellan, which might represent a barrier to dispersal. Summing up, Subpop B presents larger genetic differences with respect to the rest of subpopulations (Table 2).

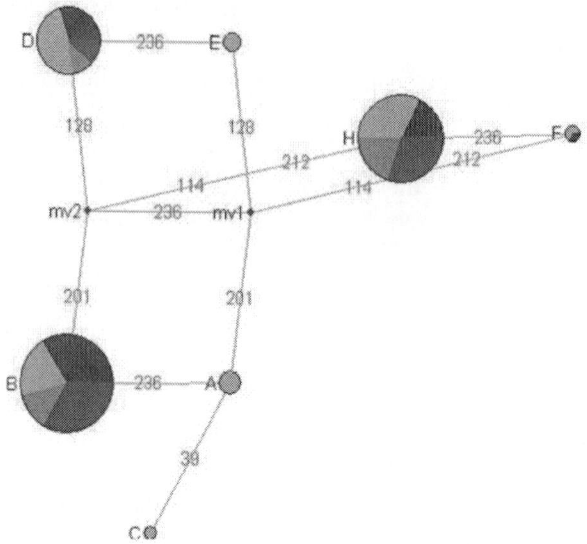

Figure 3. Haplotype network of *Castor canadensis*. Circle diameters are proportional to the number of individual sequences per haplotypes, colors indicates subpopulations (violet (subpopA), orange (subpopB), brown (subpopC), green (subpopD) and red (dubpopE)) and numbers represent nucleotide differences. Small red circles represent hypothetical haplotypes (mv1, mv2).

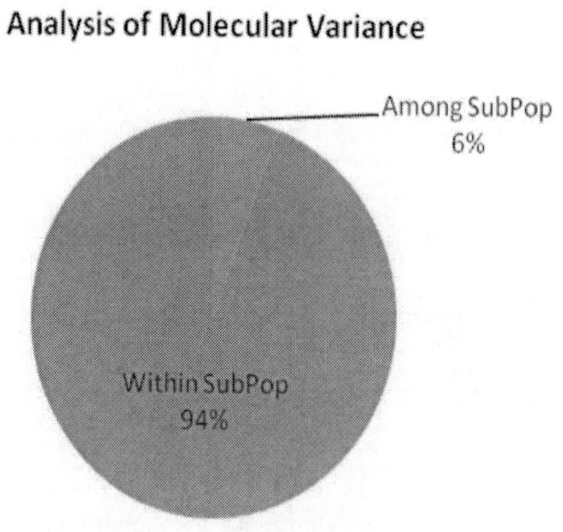

Figure 4. Results of Analysis of Molecular Variance. Percentages of variation among and within subpopulations.

Table 2. Pairwise population differentiation values between five subpopulations of *Castor canadensis*. p values are shown above diagonal, ϕst values are shown below diagonal

	SubPop A	SubPop B	SubPop C	SubPop D	SubPop E
SubPop A		0.069	0.358	0.291	0.323
SubPop B	0.032		0.001	0.001	0.007
SubPop C	0.000	0.100		0.019	0.079
SubPop D	0.003	0.082	0.053		0.354
SubPop E	0.002	0.091	0.050	0.000	

Spatial Autocorrelation

The Spatial Autocorrelation Analysis between individuals from PNTDF showed a slight but significant positive relationship between genetic and geographic distances ($r = 0.094$; $p = 0.002$; Figure 5). This pattern was also shown (albeit not significantly) for the first two distance classes (up to 1.3 km), where the correlation coefficient was positive, but non-significant, indicating that beavers 5 km apart from each other are more genetically related than those that are located at smaller and greater distances. On the other hand, beavers that are between 2 and 5 km and > 6 km apart, presented a negative, non-significant correlation coefficient.

Relationship between Northern Hemisphere and TDFA Haplotypes

Samples coming from Alaska (USA) presented a single haplotype (Haplotype G), while of a total of five samples from Minnesota (USA), four presented haplotype MN2 and the other one, haplotype MN3. All haplotypes were deposited in GenBank (Table 3).

The haplotype network (Figure 6) evidenced that the haplotypes found in TDFA are closely related, with an average nucleotide number between haplotypes of 1.873; while those haplotypes from the northern Hemisphere present a higher number of changes between haplotypes ($k = 6.733$).

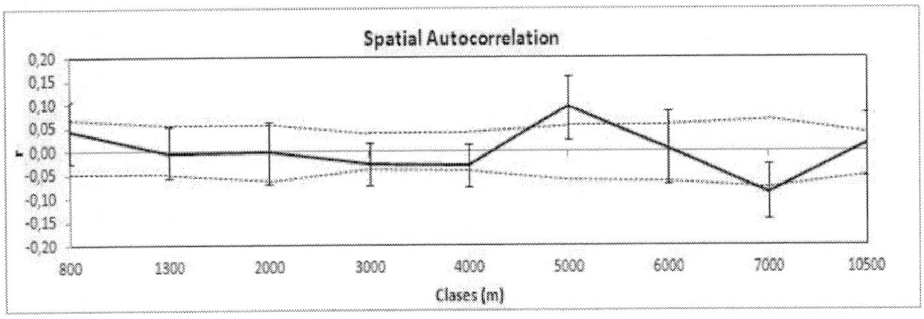

Figure 5. Spatial Autocorrelation Analysis. Spatial genetic structure autocorrelograms for the National Park. The 95% confidence intervals for the autocorrelation coefficients (r) are also shown (broken line).

Even if none of the haplotypes detected in TDFA were found in the northern Hemisphere haplotypes, it can be mentioned that haplotypes D, H and B (all haplotypes from TDFA) are closer to two haplotypes found in the northern Hemisphere (*i.e.* Haplotypes G and MN2) (Figure 6). The separation between them is three polymorphic sites (*e.g.* Hap H *vs* MN2; Hap B *vs* Hap G). Given the short time since the introduction of North American beavers to TDFA, it is unlikely that new haplotypes have arisen due to mutations, and therefore, the haplotypes found in the Archipelago must be present in the northern Hemisphere due to their origin.

DISCUSSION

Population Structure

The colonization of new areas generally produces a marked founder effect with genetic drift, which reduced the genetic variability within the populations and increases the genetic differentiation between populations (Wang *et al.* 2008). However, on the beaver population in TDF, a high intrapopulation variability (94%) was detected together with a low genetic differentiation (6%) among subpopulations. The same pattern was found by Refoufi and Esnault (2006).

According to Lizarralde *et al.* (2008), a maximum of twenty-five hypothetical mitochondrial lineages could be the founders of the invasive population in TDFA; of those, seven lineages were identified in this study.

Chances are that the twenty-five hypothetical lineages (1) might not have been present in the founding population; or (2) if present in the founders, might have disappeared or became extinct through selection during the processes of colonization, expansion and invasion.

The haplotype network showed that three haplotypes (B, D and H) were the most abundant ones, and were detected in five subpopulations, while the rest were registered in lower proportions in one or two subpopulations. Two hypothetical haplotypes were also detected, which might be considered "ancestral" given that they connect all the haplotypes and that: a) they might really exist in the population but were not included in the sampling; b) they might have been originally introduced in TDF but later became extinct; or c) they might only be present in populations from the northern Hemisphere.

The haplotype network was built including northern Hemisphere haplotypes, and one of the hypothetic haplotypes (mv2) was directly connected to these, which might indicate their ancestral nature. On the other hand, the three most abundant haplotypes (B, D and H) which are directly connected to the ancestral haplotype (mv2) only differ from this one in a base pair. The latter might indicate that haplotypes closer to those from the northern Hemisphere are the most common in TDFA. Lastly, the hypothetical haplotype (mv2) would be the most ancestral one, given that it originates six haplotypes. This hypothetical haplotype might probably be found in Alberta, Canada where the founding stock came from.

Table 3. Polymorphic sites at the control region of mitochondrial DNA for *Castor canadensis* from South America (Hap A–F and Hap H) and North America (CA1, CA2 and CA3 (Ducroz *et al.* 2005), HAPLOTYPE G, MN2, MN3). GenBank accession numbers are shown

Nucleotide Position																				
	39	114	128	144	201	205	212	223	233	236	268	276	300	353	354	401	416	420	421	422	Acceso GenBank
HAPLOTYPEA	A	C	C	T	G	T	T	C	G	G	C	G	A	T	C	C	A	C	T	C	AY787822
HAPLOTYPE B										A											AY787823
HAPLOTYPE C	G																				AY787824
HAPLOTYPE D			T		A					A											AY787825
HAPLOTYPE E			T		A																AY787826
HAPLOTYPE F		T			A		C														AY787827
HAPLOTYPE G					A			T		A											AY968083
HAPLOTYPE H		T			A		C			A											EU476079
MN2					A					A		A	G								JN655158
MN3		T		C	A			T	A	A	T			C	T	T					JN655159
CA1					A		C			A							C	T	C	T	AY623644
CA2					A	C				A		A					C	T	C	T	AY623645
CA 3					A	C				A		A					C	T	C	T	AY623646

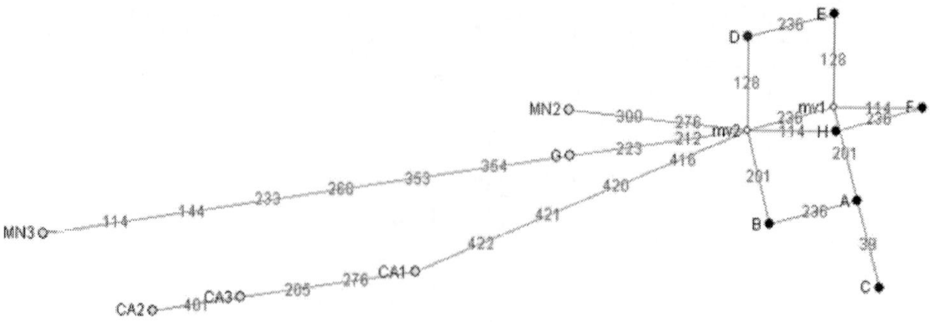

Figure 6. Haplotype network of *Castor canadensis*. In black: Tierra del Fuego haplotypes. In yellow: North Hemisphere haplotypes. The numbers represent the 20 nucleotide differences.

Assuming an island model, Wright (1978) proposed that values of Fst > 0.25 would evidence a great differentiation among populations; values between 0.15 and 0.25 would represent a moderate differentiation, while there would be no differentiation with values of Fst < 0.05. In practice, values of Fst rarely exceed 0.5; on the contrary, they are much less. The complex hydrographic network in TDFA would provide means for beavers to move freely across the region, behaving as a large, panmictic population, which is backed by the results presented here. A small but significant divergence between TDFA subpopulations was registered ($\phi_{st} = 0.058$; $p = 0.002$).

The ways in which animals make use of the landscape are determined by their habitat requirements and their social structure; therefore, populations can exhibit a spatial genetic structure even in the absence of physical barriers (Latch *et al.* 2008). However, in this work, those subpopulations that exhibited a smaller gene flow were Subpop B and C ($\phi_{st} = 0.100$; $p = 0.001$) and Subpop B and E ($\phi_{st} = 0.091$; $p = 0.007$), which are located further apart geographically. Subpopulation E is separated from B by the Strait of Magellan, which might pose a barrier to dispersal (Hoffmann 1985), although the limited gene flow might also be due to the great geographic distance (> 230 km) between both subpopulations. Subpop B and E are very different in their landscape use and density patterns. Subpopulation B inhabits a forested zone and presents a higher abundance, while subpopulation E is located on the steppe and presents the lowest density.

Finally, the Spatial Autocorrelation Analysis provided a finer-scale resolution for the population genetic structure. According to these results, beavers 5 km apart (in linear dimensions) are more genetically similar among themselves than those that are separated by smaller and larger distances. This might indicate that juveniles migrating from a source colony to start their own colonies, might do so within a 5 km radius.

Management Strategies for the North American Beaver in TDF

As ecosystem engineers and keystone species, beavers have great impacts on the ecosystems and biodiversity of TDFA, and on the local economies. The North American beaver had a surprising population expansion shortly after its introduction, which brought large-scale environmental changes. Deciding upon its eradication, a control strategy or even if the species should be tolerated, is no doubt a complex choice.

Firstly, the North American beaver has long become a successfully-established invasive species in TDFA, partly due to the absence of predators and competitors, favorable environmental conditions, the inaccessibility of many remote areas where it thrives, and most importantly, the complex hydrographic network that facilitates the invasion. In turn, all of these factors considered together call for any management action to be decided upon multiple environmental and social aspects. For instance, it is largely unknown what potential effects the eradication of the North American beaver would have over the structure and functioning of current TDFA ecosystems. Some topics to be considered firsthand include the effects of the breakdown of dams and flooding of large forest areas, the redistribution of the sediments trapped in dams and their downstream transport to the ocean, the ecological restoration of impacted areas and their recovery times in terms of biogeochemical cycles. Worldwide, only a few mammal eradication programs have been successful, and almost exclusively, they took place in islands or small populations (Towns and Broome 2003; Clout and Rusell 2006; Martins *et al.* 2006; Rodríguez *et al.* 2006). Therefore, just considering an eradication strategy for the North American beaver in Isla Grande (with an area of ~ 48,000 km^2), might represent an effort two-orders of magnitude above any successful rodent eradication program implemented to date (*e.g.* 110 km^2; Towns and Broome 2003), and probably the least convenient alternative from a logistics point of view. Technical difficulties (*e.g.* environment, climate, inaccessibility to invaded sites, a complex hydrographic network) would prevent the extraction of 100% of individuals. Moreover, financial constraints related to the very high costs of eradication (estimated in US $30 million; Parkes *et al.* 2008) and its estimated long period of implementation (over 20 years; Parkes *et al.* 2008), would induce managers to decide against it.

Social aspects should also be considered together with the above-mentioned problems. As the invasion progressed, beavers paradoxically become "emblem" animals for the tourism industry and for the inhabitants of TDFA (Parkes *et al.* 2008), and so the community's opinion on management matters should be taken into account. On the other hand, for an eradication program to be successful, the detection of (natural or artificial) barriers to dispersion is paramount, so that each discrete population (< 110 km^2) could be considered an Eradication Unit. However, results from this study indicate that it is not possible to clearly identify EU in Isla Grande, given that the Strait of Magellan would be the only geographic barrier that would prevent gene flow in the population. The scenario then becomes complex, and makes it difficult to decide whether eradicating, controlling or even tolerating the species is the more convenient strategy, and calls for a combination of all alternative measures into a single strategy.

It has been shown that the TDFA population does not present a marked spatial structure, while the five subpopulations identified at the molecular and demographic levels have smaller size and therefore would be easier to manage. Therefore, using diagnostic tools from molecular biology techniques, five management units (MU) can be defined, which are representative of the "typical" environments invaded by beavers. Two types of activities are suggested:

1. Eradication in basins or pilot water streams selected for each MU, and
2. Control of the expansion to other occupied areas.

The objective would then be to *eradicate* individuals in these *MU*, in accordance to the technical and economic resources available, and to generate the missing scientific information that would enable an objective analysis of the eradication in TDFA. For each MU, sustained capture efforts would be put into practice and they would be considered "Eradication Units". In that sense, the "best practices" would be used based on previous studies (Lizarralde and Escobar 2000) sustaining superior capture efforts than those reported for the management plan. Lizarralde and Escobar (1999). *Control* strategies would emphasize and stimulate extraction by poachers and independent hunters, together with local inhabitants especially in forested areas, promoting a renewed interest in sound poaching practices, with a basis in environmental conservation. These measures would tend to lower the pressure exerted by the species and its re-invasion of MU. The ultimate control goal would be to effectively poach all beavers from all *colonies*. Accordingly, the "best practices", which indicate poaching efforts of 30,000 to 38,000 traps per night (Lizarralde and Escobar 1999; Lizarralde 2008), would be carried out. The extraction could be further promoted with a renewed interest in the exploitation of beaver fur and meat (Garriz and Lizarralde 2007), therefore generating an increased participation of public and private entities in the control of the exotic species and indirectly contributing to the design of economically-viable management plans. In this way, an economic income could be instated for a region that lacks any major economic developments of its natural resources and that would generate funds that could be destined for the conservation of the regional biodiversity.

The recent invasion of a continental section of Chile (Brunswick Peninsula) forced management agencies from both countries to sign a bi-national agreement in 2008 for the restoration of those ecosystems affected by beavers. Beaver invasion is no doubt a complex topic, and a molecular biology approach to it would allow to amalgamate different strategies in order to ensure a successful eradication and control program that works in the short- and mid-terms, in which the scientific community, private interests and non-government organizations would participate.

A sustainable management unit that includes the criteria outlined in this chapter, is likely to alleviate economic losses due to reduced wood production and flooding of foraging grounds. Moreover, it would contribute to the restoration of *Nothofagus* forests, helping restore specific diversity in the habitat currently occupied by the North American beaver. Lastly, an effective control plan would reduce dispersal of numerous exotic plant species that are associated to the habitats modified by beavers.

ACKNOWLEDGMENTS

We thank CONICET for financial support (Grants PIP 0123/2008 and iBOL project) and also Universidad Nacional de La Plata for providing services. Special thanks to Secretaría de Ambiente y Recursos Naturales de TDF (Argentina), Parque Nacional TDF (Argentina), Servicio Agrícola Ganadero (Mr. Nicolás Soto, SAG, Chile), Mr. Derek Corcorán (Universidad Católica de Chile), and Mr. Claudio Moraga (Wildlife Conservation Society, WCS, Parque Karukinka). Samples from Northern hemisphere populations were made available by Dr. Thomas Hanley (US Forest Service, Juneau, Alaska, USA), and Dr. Karla Pelz-Serrano (School of Natural Resources, University of Arizona, USA).

REFERENCES

Administración de Parques Nacionales (APN). (2006). Programa de Control de "Castor Americano" *Castor canadensis*. 20 pp.

Aljanabi, S. M.and Martínez, E. I. (1997). Universal and rapid salt-extraction of high quality genomic DNA for PCR- based techniques. *Nucleic Acids Research*. 25:4692-4693.

Anderson, C. B., Griffith, C. R., Rosemond, A. D., Rozzi, R., and Dollenz, O. (2006). The effects of invasive North American beavers on riparian plant communities in Cape Horn, Chile - Do exotic beavers engineer differently in sub-Antarctic ecosystems? *Biological Conservation* 128(4): 467-474.

Anderson, C. B., Martínez Pastur, G., Lencinas, M. V., Wallem, P. K., Moorman, M. C., and Rosemond, A. D. (2009). Do introduced North American beavers engineer differently in southern South America? An overview with implications for restoration. *Mammal Review*. 39:33–52.

Anderson, C. H., Soto, N., Cabello, J. L., Martínez Pastur, G., Lencinas, M. V., Wallem, P., Antúnez, D., and Davis, E. (2011). Building effective alliances between research and management to mitigate the impacts of an invasive ecosystem engineer: Lessons from the pioneering example of the study and control of North American beavers in the Fuegian archipelago. In: R. Francis (Ed.). *A handbook of global freshwater invasive species*. Earthscan Press, London, UK. Section 5, Chapter 29, pp 347-359 (ISBN 978-1-849-71228-6).

Andrade, B. S. (2005). Gestión para la caza y aprovechamiento de fauna dañina en XII Región. *Informe preparado para el Servicio Agrícola y Ganadero de Magallanes*, 107 pp.

Barnes, D. M. and Mallik, A. U. (1997). Habitat factors influencing beaver dam establishment in a northern Ontario watershed. *Journal of Wildlife Management*. 61:1371-1377.

Briones, M., Schlatter, R., Wolodarsky, A. Andvenegas, C. (2001). Clasificación ambiental para hábitat de *Castor canadensis* (Kuhl 1820, Rodentia), de acuerdo a características de cuencas en un sector de Tierra del Fuego. *Anales del Instituto de la Patagonia*, 29, 75–93.

Busher, P. E. (1996). Food catching behavior of beavers (Castor canadensis): selection and use of woody species. *American Midland Naturalist*. 135: 343-348.

Castillo, N. (2006). Determinación de la composición botánica de la dieta de Castor (*Castor canadensis*) en la isla Tierra del Fuego. Tesis título de Ingeniero en Ejecución Agropecuario. *Universidad de Magallanes, Punta Arenas*.

Clout, M. N. and Rusell, J. C. (2006). The eradication of mammals from New Zealand islands. In: Koike, F., Clout, M. N., Kawamichi, M., De Poorter, M., and Iwatsuki, K. (Eds.). *Assesment and Control of Biological Invasion Risks*. (Pp. 127-141). Shoukadoh Book Sellers, Kyoto, Japan and UICN, Gland, Switzerland.

Collen, P. and Gibson, R. J. (2001). The general ecology of beavers (*Castor spp.*) as related to their influence on stream ecosystems and riparian habitats, and the subsequent effects on fish – a review. *Review in Fish Biology and Fisheries* 10: 439-461.

Curtis, P. D. and Jensen, P. G. (2004). Habitat factors affecting beaver occupancy along roadsides in New York State. *Journal of Wildlife Management*. 68(2):278-287.

Ducroz, J. F., Stubbe, M., Saveljev, A. P., Heidecke, D., Samjaa, R., Ulevicius, A., Stubbe, A., and Durka, W. (2005). Genetic variation and population structure or the Eurasian Beaver *Castor fiber* in Eastern Europe and Asia. *Journal of Mammalogy*. 86(6): 1059-1067.

Durka, W., Babik, W., Ducroz, J. F., Heidecke, D., Rossel, F., Samjaa, R., Saveljev, A. P., Stubbe, A., Ulevicius, A., and Stubbe, M. (2005). Mitochondrial phylogeography of the Eurasian beaver *Castor fiber* L. *Molecular Ecology*. 14: 3843-3856.

Elton, C. S. (1958). *The Ecology of Invasions by Animals and Plants*. Methuen and Co., London, UK.

Fasanella, M., Poljak, S. and Lizarralde, M. (2010). Invasive North American Beaver (*Castor canadensis*): the distribution of mitochondrial variation across the Archipelago of Tierra del Fuego. *Mastozoología Neotropical*. 17(1):43-52.

Fluxus Technology LTD. (1999-2011). NETWORK ver 4.6.0.0. Available from http://www.fluxus-engineering.com.

Garriz, C. and Lizarralde, M. (2007). Subproductos Cárnicos de Fauna Silvestre: Posibilidad de aprovechamiento de carne de castor. *Publicación Científico Técnica Recursos Naturales* (Formato Digital). SPD Provincia de Tierra del Fuego. 70 pp.

Hall, T. A. (1999). BioEdit: a user-friendly biological sequence alignment editor and analysis program for Windows 95/98/NT. *Nucleic Acids Symposium Series*. 41:95-98.

Hartman, G. (1995). Patterns of spread of a reintroduced beaver Castor fiber population in Sweden. *Wildlife Biology*. 1:97-103.

Hoffmann, M. (1985). Zur Ansiedlung der Bisamratte,Ondatra zibethica (L.) In: Südamerika (On the settlement of the muskrat, *Ondatra zibethica* (L.) in South-America). *Journal Pest of Science*. 58: 93-97.

Iriarte, A. (2000). Impacto de la fauna silvestre sobre la flora nativa de Chile. In: Baldini, A. and Pancel, L. (Eds). *Agentes de daño en el bosque nativo.* Santiago, Chile. Editorial Universitaria. 319-350.

Jaksic, F. M., Yánez, J. L. and Rau, J. R. (1983). Trophic relations of the southernmost population of Dusicyon in Chile. *Journal of Mammalogy* 64: 693-697.

Kolar, C. S. and Lodge, D. M. (2001). Progress in invasion biology: predicting invaders. *Trends in Ecology and Evolution*. 16(4): 199-204.

Latch, E. K., Scognamillo, D. G., Fike, J. A., Chamberlain ,M. B., and Rhodes, O. E. (2008). Deciphering ecological barriers to North American River Otter (*Lontra canadensis*) gene flow in the Louisiana landscape. *Journal of Heredity* 99(3): 265-274.

Lever, C. (1994). *Naturalized Animals: the ecology of successfully introduced species.* T. and A. D. Poyser, London. 354 pp.

Lizarralde, M. (1989). El Castor (*Castor canadensis*) en Tierra del Fuego. *Praxis* 1: 34-35.

Lizarralde, M. (1993). Current status of the introduced beaver (*Castor canadensis*) population in Tierra del Fuego, Argentina. *Ambio* 22(6): 351-358.

Lizarralde, M. S., Deferrari, G. A., Alvarez, S. E., and Escobar, J. M. (1996). Effects of beaver (*Castor canadensis*) on the nutrient dynamics of the Southern Beech forest of Tierra del Fuego (Argentina). *Ecología Austral*. 6: 101-105.

Lizarralde, M. and Escobar, J. (1999). Plan de Manejo de la especie *Castor canadensis*. *Sub. De Rec. Naturales, Prov*. TDF.

Lizarralde, M. and Escobar, J. M. (2000). Mamíferos exóticos en la Tierra del Fuego. *Ciencia Hoy* 10(56): 52-63.

Lizarralde, M., Escobar, J. and Deferrari, G. (2004). Invader species in Argentina: A review about the beaver (Castor canadensis) population situation on Tierra del Fuego ecosystem. *Interciencia* 29(7): 352-356.

Lizarralde, M. S., Bailliet, G., Poljak, S., Fasanella, M., and Giulivi, C. (2008). Assessing genetic variation and population structure of invasive North American beavers (*Castor canadensis* Kuhl, 1820) in Tierra del Fuego (Argentina). *Biological Invasions* 10: 673-683.

Martínez Pastur, G., Lencinas, M. V., Escobar, J., Quiroga, P., Malmierca, L., and Lizarralde, M.. (2006). Understory succession in Nothofagus forests affected by *Castor canadensis* in Tierra del Fuego (Argentina). *Applied Vegetation Science* 9: 143-154.

Martins, T. L. F., Brooke, M. L., Hilton, G. M., Farnsworth, S., Gould, J., and Pain, D .J. (2006). Costing eradications of alien mammals from islands. *Animal Conservation*. 9:439-444.

Massoia, E. and Chebez, J. C. (1993). *Mamíferos silvestres del archipiélago fueguino*. Editorial L.O.L.A., Buenos Aires.

Moorman, M. C., Anderson, C. B., Gutiérrez, A. G., Charlin, R., and Rozzi, R. (2006). Watershed conservation and aquatic benthic macroinvertebrate diversity in the Alberto D'Agostini National Park, Tierra del Fuego, Chile. *Anales del Instituto de la Patagonia*. 34, 41–58.

Moritz, C. (1994). Defining "Evolutionary significant units" for conservation. *Trends in Ecology and Evolution*. 9:373-375.

Müller-Schwarze, D. and Sun, L. (2003). *The Beaver. Natural History of a Wetlands Engineer*. Cornell University Press, Ithaca, New York.

Naiman, R. J., Johnston, C. A. and Kelley, C. A. (1988). Alteration of North American streams by beaver: The structure and dynamics of streams are changing as beaver recolonize their historic habitat. *BioScience* 38:753-762.

Palsbøll, P. J., Berubé, M. and Allendorf, F. W. (2006). Identification of management units using population genetic data. *Trends in Ecology and Evolution* 13:146-158.

Parkes, J. P., Paulson, J., Donlan, C. J., and Campbell, K. (2008). Control of North American Beavers in Tierra del Fuego: Feasibility of Eradication and Alternativa Management Options. *Landcare Research Contract Report* LC0708.

Peakall, R. and Smouse, P. E. (2006). GENAIEX 6: genetic analysis in Excel. Population genetic software for teaching and research. *Molecular Ecology Notes*. 6: 288-295.

Pimentel, D., Lach, L., Zúñiga, R., and Morrison, D. (2000). Environmental and economic costs of non-indigenous species in the United States. *BioScience*. 50, 53–65.

Ramírez Silva, M. D. (2006). Cuantificación de la biomasa leñosa removida por *Castor canadensis* (Kuhl 1820, Rodentia) en bosques nativos de Tierra del Fuego (XII región de Magallanes, Chile). *Tesis de doctorado, Departamento de Ciencias Forestales*, Pontificia Universidad Católica de Chile, 152 pp.

Refoufi, A. and Esnault, M. A. (2006). Genetic diversity and population structure of Elytrigia pycnantha (Godr.) (Triticeae) in Mont Saint-Michel bay using microsatellite markers. *Plant Biology*. 8:234-242.

Robertson, B. C. and Gemmell, N. J. (2004). Defining eradication units to control invasive pests. *Journal of Applied Ecology*. 41:1042-1048.

Rodríguez, C., Torres, R. and Drummond, H. (2006). Eradicating introduced mammals from a forested tropical island. *Biological Conservation*. 130:98-105.

Rozas, J., Sánchez-Del Barrio, J. C., Messeguer, X., and Rozas, R. (2003). DNAsp. DNA polymorphism analyses by the coalescent and other methods. *Bioinformatics*. 19:2496-2497.

Sala, O. E., Chapin, F. S., Armesto, J. J., Berlow, E., Bloomfield, J., Dirzo, R., Huber-Sannwald, E., Huenneke, L. F., Jackson, R. B., Kinzig, A., Leemans, R., Lodge, D. M., Mooney, H. A., Oesterheld, M., Leroy, Poff, N., Sykes, M. T., Walker, B. H., Walker, M., and Wall, D. H. (2000). Global biodiversity scenarios for the year 2100. *Science*. 287: 1770-1774.

Sambroock, J., Fritsch, E. F. and Maniatis, T. (1989). *Molecular cloning, a Laboratory Manual*. 2nd Edition. Cold Spring Harbor Laboratory Press.

Silva, C. A. and Saavedra, B. (2008). Actas del Taller Internacional para el Control de Castores en la Patagonia. Edición digital. *Wildlife Conservation Society - Chile*.

Skewes, O., González, F., Rubilar, L., Quezada, M., Olave, R., Vargas, V., and Ávila, A. (1999). Investigación, aprovechamiento y control de castor (*Castor canadensis*) en las islas Tierra del Fuego y Navarino. Informe Final. Servicio Agrícola y Ganadero (SAG) XII Región, *Magallanes y Antártica chilena*. 200 pp.

Skewes, O., Gonzalez, F., Olave, R., Ávila, A., Vargas, V., Paulsen, P., König, H. E. (2006). Abundance and distribution of American beaver, *Castor canadensis* (Kuhl 1820), in Tierra del Fuego and Navarino islands, Chile. *European Journal of Wildlife Research*. 52: 292–296.

Suzuki, N. and McComb, W. (1998). Habitat classification models for Beaver (*Castor canadensis*) in the streams of the central Oregon coast range. *Northwest Science*. 72:102-110.

Taylor, R. H., G. W. Kaiser and M. C. Drever. (2000). Eradication of Norway rats for recovery of seabird habitat on Langara Island, Britsh Columbia. *Restoration Ecology*. 8:151-160.

Towns, D. R. and Broome, K.,G. (2003). From small Maria to massive Campbell: forty years of rat eradications from New Zealand islands. *New Zealand Journal of Zoology*. 30: 377-398.

Vázquez, D. P. (2002). Multiple effects of introduced mammalian herbivores in a temperate forest. *Biological Invasions*. 4: 175-191.

Vilà, C., Amorim, I. R., Leonard, J. A., Posadas, D., Castroviejo, J., Petrucci-Fonseca, F., Crandall, K. A., Ellegren, H., and Wayne, R. K. (1999). Mitochondrial DNA phylogeography and population history of the grey wolf *Canis lupus*. *Molecular Ecology*. 8: 2089-2103.

Wallem, P. K., Jones, C. G., Marquet, P. A., and Jaksic, F. M. (2007). Identificación de los mecanismos subyacentes a la invasión de *Castor canadensis* (Kuhl 1820, Rodentia) en el archipiélago de Tierra del Fuego, Chile. *Revista Chilena de Historia Natural*. 80: 309-325.

Wang, T., Su, Y. and Chen, G. (2008). Population genetic variation and structure of the invasive weed *Mikania micrantha* in Southern China: consequences of rapid range expansion. *Journal of Heredity*. 99(1):22-33.

Wright, J. P., Jones, C. G. and Flecker, A. S. (2002). An ecosystem engineer, the beaver, increases species richness at the landscape scale. *Oecologia*. 132(1):96-101.

Wright, S. (1978). Evolution and the Genetics of Populations. In: *Variability Within and Among Natural Populations*. Vol. 4. University of Chicago Press, Chicago.

In: Invasive Species
Editors: Joaquín J. Blanco and Adrian T. Fernandes
ISBN: 978-1-61942-761-7

Chapter 5

LEAF CUTTING-ANTS: CONTROL METHODS AND SUSTAINABILITY

Gabriela Ortiz and Maria Izabel Camargo Mathias [1]
Universidade Estadual Paulista Júlio de Mesquita filho (UNESP),
Instituto de Biociências, Rio Claro/SP, Brasil

ABSTRACT

The fungus-growing ants belong to the genus Atta and Acromyrmex and individuals are causing serious damage to various types of plantations due to the habit of cutting leaves and shoots of economic importance for use in cultivating the garden symbiotic fungus colonies, causing serious damage. The ants attack is harmful at any stage of the cycle, but the damage is greater in the training phase of the plant, when temporarily paralyzes growth. In an attempt to reduce the damage caused by these insects farmers make the control of fire ants through the use of mechanical methods, biological and chemical. The mechanical control is to dig the nest to remove the queen (individual responsible for placing the eggs that give rise to other individuals in the colony), which are larger in size than workers, causing a disruption of the nest, since the queen controls all activities and individuals in the colony. This type of control has been recommended only for newly formed nests, otherwise, the queen will be housed in great depth, making its removal. It should be noted that in practice such control is not a good alternative in commercial plantations due to the large extent of the property. Chemical control can be accomplished by applying powder, liquid insecticide, thermonebulization and granulated bait. The powder is a type of insecticides formulated in solid vehicles, such as talc, and the application is through hand pumps. The death of ants occurs by direct contact with the chemical. However, this type of control has limitations such as the impossibility of penetration of the product in the deepest chambers of nesting adults, and has now been replaced by liquid thermonebulized, through which the product is injected into the nest with the help of hose, and toxic bait, which is considered more efficient due to easy handling, fast action and lower costs. The bait is distributed on the tracks near the colony of ants to be transported into the nest by workers. This method can cause much damage to non-target ants and even other bodies. Finally, it has been used the biological control,

---

[1] Email: *gaortiz@rc.unesp.br*.

which is the introduction of natural enemies in plantations, such as birds, some species of spiders, beetles, mites and even other species of ants, and much research is currently focused on creating natural insecticides that have repellent action or lethal to the ants, such as castor oil, or fungicidal against the garden of fungus that feeds the colony, as is the case of sesame. Studies on the morphology of the reproductive organs of these ants could also be an important tool in the search for new methods of control. Some studies have been conducted on the morphology of the spermathecae of ants, the body that produces all the elements capable of maintaining viable sperm inside, where it was observed that products circulating in the hemolymph, due to the thin wall thickness of the the spermatheca, also could enter the environment within this structure, where are the sperm and thus affect their viability.

INTRODUCTION

Leaf cutting ants, a non-generic name, are any of 47 species (Speight et al., 1999) of leaf-chewing ants belonging to the two genera *Atta* and *Acromyrmex*. These species of tropical, fungus-growing ants are all endemic to South, Central America, Mexico and parts of the southern United States. The *Acromyrmex* and *Atta* ants have much in common anatomically; however, the two can be identified by their external differences. *Atta* ants have three pairs of spines and a smooth exoskeleton on the upper surface of the thorax, while *Acromyrmex* ants have four pairs and a rough exoskeleton (Weber, 1972).

Next to humans, leafcutter ants form the largest and most complex animal societies on Earth. In a few years, the central mound of their underground nests can grow to more than 30 meters (98 ft) across, with smaller, radiating mounds extending out to a radius of 80 meters (260 ft), taking up 30 to 600 square meters (320 to 6,500 sq ft) and containing eight million individuals (Speight et al., 1999).

Winged females and males leave their respective nests *en masse* and engage in a nuptial flight known as the *revoada*. Each female mates with multiple males to collect the 300 million sperm she needs to set up a colony (Piper, 2007).

Once on the ground, the female loses her wings and searches for a suitable underground lair in which to found her colony. The success rate of these young queens is very low, and only 2.5% will go on to establish a long-lived colony. To start her own fungus garden, the queen stores bits of the parental fungus garden mycelium in her infrabuccal pocket, which is located within her oral cavity (Weber, 1972).

In a mature leafcutter colony, ants are divided into castes, based mostly on size, that perform different functions. *Acromyrmex* and *Atta* exhibit a high degree of biological polymorphism, four castes being present in established colonies - minims, minors, median and majors. Majors are also known as soldiers or dinergates. *Atta* ants are more polymorphic than *Acromyrmex*, meaning there is comparatively less difference in size from the smallest to largest types of *Acromyrmex*.

Minims are the smallest workers, and tend to the growing brood or care for the fungus gardens. Head width is less than 1mm. Minors are slightly larger than minima workers, and are present in large numbers in and around foraging columns. These ants are the first line of defense and continuously patrol the surrounding terrain and vigorously attack any enemies that threaten the foraging lines. Head width is around 1.8-2.2mm. Median is the generalized foragers, which cut leaves and bring the leaf fragments back to the nest. Majors, the largest worker ants, act as soldiers, defending the nest from intruders, although there is recent evidence that majors participate in other activities, such as clearing the main foraging trails of large debris and carrying bulky items back to the nest. The largest soldiers (*Atta laevigata*) may have total body lengths up to 16mm and head widths of 7mm.

Leaf-cutter ants have very specific roles when it comes to taking care of the fungal garden and dumping the refuse. Waste management is a key role for each colony's longevity. The necrotrophic parasite *Escovopsis* of the fungal cultivar threatens the ants' food source, and is thus a constant danger to the ants. The waste-transporters and waste heap workers are the older, more dispensable leaf-cutter ants, ensuring the healthier and younger leaf-cutter ants can work on the fungal garden. The *Atta colombica* species, unusually for the *Attine* tribe, have an external waste heap. Waste-transporters take the waste, which consists of used substrate and discarded fungus, to the waste heap. Once dropped off at the refuse dump, heap-workers organize the waste and constantly shuffle it around to aid decomposition. A compelling observation of *Atta colombica* was that the dead ants were placed around the perimeter of the waste heap (Hart and Ratnieks, 2002; Bot, 2001).

In addition to feeding the fungal garden with foraged food, mainly consisting of leaves, it is protected from *Escovopsis* by the antibiotic secretions of *Actinobacteria* (genus *Pseudonocardia*). This mutualistic micro-organism lives in the metapleural glands of the ant *Actinobacteria* are responsible for producing the majority of the world's antibiotics today (Zhang, 2007).

When the ants are out collecting leaves, they are at risk of being attacked by the phorid fly, a parasitic pest which lays eggs into the crevices of the worker ants' head. Often a minim will sit on the worker ant and ward off any attack. Also, the wrong type of fungus can grow during cultivation. Escovopsis is a highly virulent fungus that has the potential to devastate an ant garden; as it is horizontally transmitted. It Currie et al. (1999) found that Escovopsis was cultured, during colony foundation, in 6.6% of colonies. However, in one to two year old colonies, almost 60% had Escovopsis growing in the fungal garden (Reynolds et al., 2004).

In some parts of their range, leafcutter ants can be quite a nuisance to humans, defoliating crops and damaging roads and farmland with their nest-making activities. For example, some Atta species are capable of defoliating an entire citrus tree in less than 24 hours. Deterring the leafcutter ant Acromyrmex lobicornis from defoliating crops has been found simpler than first expected. Collecting the refuse from the nest and placing it over seedlings or around crops resulted in a deterrent effect over a period of 30 days (Ballari and Farji-Brener, 2006).

The fungus-growing ants belong to the genus Atta and Acromyrmex and individuals are causing serious damage to various types of plantations due to the habit of cutting leaves and shoots of economic importance for use in cultivating the garden symbiotic fungus colonies, causing serious damage.

The ants attack is harmful at any stage of the cycle, but the damage is greater in the training phase of the plant, when temporarily paralyzes growth. In an attempt to reduce the damage caused by these insects farmers make the control of ants through the use of mechanical methods, biological and chemical (Ballari and Farji-Brener, 2006).

The mechanical control is to dig the nest to remove the queen (individual responsible for placing the eggs that give rise to other individuals in the colony), which are larger in size than workers, causing a disruption of the nest, since the queen controls all activities and individuals in the colony. This type of control has been recommended only for newly formed nests, otherwise, the queen will be housed in great depth, making its removal. It should be noted that in practice such control is not a good alternative in commercial plantations due to the large extent of the property (Ballari and Farji-Brener, 2006).

Chemical control can be accomplished by applying powder, liquid insecticide, thermonebulization and granulated bait. The liquid insecticides, widely used in the past, were quickly replaced with bait or thermonebulization. This is because the products tested had low efficiency caused by the need of liquids come in contact with the ants, and the expensive work of perforation of the nest and losses by soil absorption.

Another form of control is the use of toxic fumigants, one of pioneering methods to combat these insects. Methyl bromide fumigant is the only product allowed for this purpose. In However, although highly effective, a high cost, requires labor- specialized work for your application and is highly dangerous for the applicator. The product, according to the researcher, is marketed under the liquid form ready for use in packaging, eliminating the use of equipment.

The powder is a type of insecticides formulated in solid vehicles, such as talc, and the application is through hand pumps. The death of ants occurs by direct contact with the chemical. However, this type of control has limitations such as the impossibility of penetration of the product in the deepest chambers of nesting adults, and has now been replaced by liquid thermonebulized, through which the product is injected into the nest with the help of hose.

The method of control with toxic baits is the most efficient and least expensive among all others. The baits are formed by a mixture of a substrate with an attractive principle toxic asset in the form of pellets. These pellets are distributed on the trails, near the colony, and transported to the interior of the nest by their own ants. It is the most common technique to control ants. It method is efficient, practical and economical, but has limitations as the inability to use during rainy periods and in very extensive. The active substrate is dried citrus pulp, although other materials have been used, such as corn, eucalyptus leaf, cassava flour, soybean meal, cane molasses and bagasse. The insecticide is usually dissolved in refined soybean oil, and later incorporated into the substrate. The formulated insecticide-bait toxic if swallowed and should act to present some features, among them the toxic action delayed, with a mortality less than 15% after the first day and greater than 85% after the fourteenth day from the offering of the baits. It should also be lethal at low concentrations, odorless, non-repellent and non- environmental damage. Until 1992, there was only one active ingredient on the market, dodecachlorine, a group product of organochlorines, so persistent in soil and bioaccumulative and on suspicion of causing cancer. Currently, two active ingredients slow-acting and intake are being used success in toxic baits: fipronil and sulfluramide. The active ingredient must act slowly, to be distributed in every pot of fungus, which be chewed and passed between the ants, reaching all the vines of tingling. There are currently marketed in Brazil about 12 000 tons per year of ant baits. According Forti, there are some popular misconceptions regarding toxic baits. Many people say that because the baits kill vent gas, which is not truth. When carrying the bait, the ants "lick" pellets and ingest particles with small amounts of insecticide, occurring poisoning by ingestion, and probably by a small contact action. The pellets, after hydrated inside the nest, fragment, enabling operators to incorporate these pieces after six hours offered. It was shown that 50% to 70% of the workers of the colony are contaminated with the insecticide within 24 hours. With the death of workers, there is a general disorganization of the culture of the fungus, allowing the growth of contaminating fungi parasites, causing the colony to death in a few days. The months most suitable for the placement of bait are between April and September, although they may be used at other times of year. As a material is very dehydrated, if applied to wet soil will hydrate, and get rid of the ants will not take it. But day after a heavy rain, with no forecast of more rainfall, it can be applied and also works (Ballari and Farji-Brener, 2006).

The calculation of the area to see the amount of insecticide to be applied is also very important. In general, the area is calculated by measuring the greatest width at greater length, and then multiplying the two measures. A colony of ants with 10 meters long and 5 meters width has an area of 50 square meters. If the manufacturer recommends 10 grams of bait insecticide for each square meter, the amount of bait to be put is 500 grams. The bait should never be applied at lower dose than recommended and should be placed beside the tracks active, without interrupt the flow of ants. They must be charged by ants, never placed directly on the scouts. The ant bait should not be stored with other chemicals not to undermine its attractiveness. Its application, like any other insecticide, requires care in handling. The applicator should wear boots or boots, PVC gloves, long pants and shirt and dust masks (Ballari and Farji-Brener, 2006).

Finally, it has been used the biological control, which is the introduction of natural enemies in plantations, such as birds, some species of spiders, beetles, mites and even other species of ants, and much research is currently focused on creating natural insecticides that have repellent action or lethal to the ants, such as castor oil, or fungicidal against the garden of fungus that feeds the colony, as is the case of sesame.

Studies on the morphology of the reproductive organs of these ants could also be an important tool in the search for new methods of control. Some studies have been conducted on the morphology of the spermathecae of ants, the body that produces all the elements capable of maintaining viable sperm inside, where it was observed that products circulating in the hemolymph, due to the thin wall thickness of the the spermatheca, also could enter the environment within this structure, where are the sperm and thus affect their viability. Leaf-cutting ants are the main agricultural pests of the neotropical region and their control can be performed through the use of chemical substances of low specificity that accumulate in the environment.

These ants are represented by some species of the genera Atta and Acromyrmex, which are the most derived of the tribe Attini and whose colonizing success of agricultural crops is attributed to their great voracity and capability to proliferate.

Ant queens live unusually long for insects, which generally have short lives. They store viable spermatozoa for a decade or more, and produce workers and reproductives over such long periods, that almost certainly has contributed to the propensity for ants to evolve diverse life histories that been so successful in a wide range of environments. After mating, which occurs in a few minutes during the mating flight, a queen is capable of storing viable spermatozoa for dozens of years; these spermatozoa will be gradually used to fertilize the eggs and thus originate all the individuals of the colony.

The ability to store spermatozoa releases from the risk and energy expenditure involved in finding subsequent mates (Thornhill and Alcock, 1983). Maintenance of spermatozoa viability can be a critical aspect of female reproduction in insects, yet the mechanisms mediating this phenomenon are poorly understood.

The maintenance of these spermatozoa for such a long time requires the presence of a specialized organ in the queens, known as spermatheca which presents a particular organization for each phylogenetic group that has been studied.

The spermatheca is an organ of ectodermic origin and it is internally coated with cuticle. It basically consists of a storage sac, often accompanied by associated glands, known as spermathecal glands, which produce secretions that allow the maintenance of a microenvironment of appropriate pH, ideal ionic constitution, and nutrients, in addition to other factors necessary for the survival of the spermatozoa. In some insects these glands might be absent and the internal coating epithelium performs the secretory role (Davey, 1985).

Generally, the spermatheca consists of a single sac, but two have been observed in Blaps (Coleoptera) and Phlebotomus (Diptera), and three occur in the most derived flies (Chapman, 1998). In ants in general only one spermatheca has been described.

The basic functions of the spermathecal secretions have been poorly studied, however, some observations suggest that their main function would be to maintain the viability of the spermatozoa during the time of mating until the fertilization of the eggs (Berry, 1985). The spermatozoa of insects become active inside the spermatheca, and may remain there for several months or up to several years.

In view of the all the facts described above and given the economical importance of these insects, the present work had the object of carrying out a study of the spermatheca of queen ants of Atta laevigata, Atta sexdens sexdens, Atta sexdens piriventris, Acromyrmex coronatus and Cyphomyrmex muelleri, through the use of the techniques of Scanning Electron Microscopy (SEM), histology, and histochemistry.

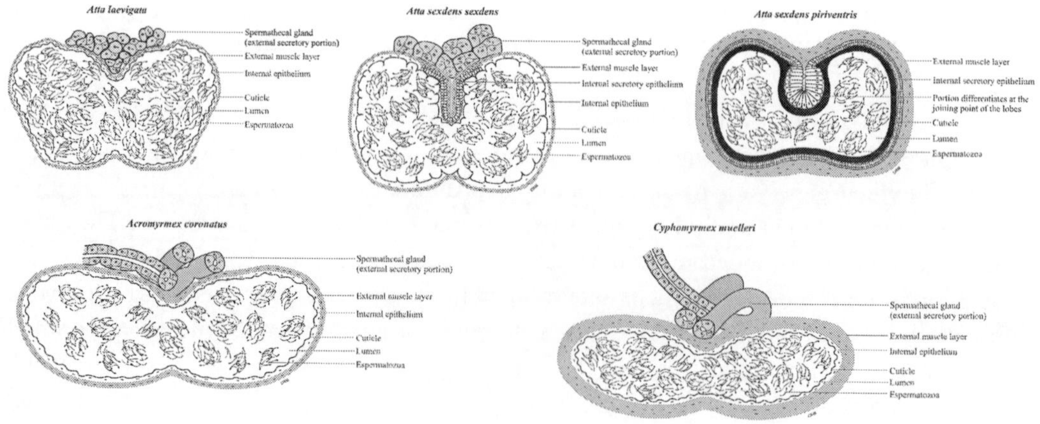

Figure.ureure. 1 . Schematic diagram showing the bilobed sac morphology of the spermatheca of the ants Atta laevigata, Atta sexdens sexdens, Atta sexdens piriventris, Acromyrmex coronatus and Cyphomyrmex muelleri.

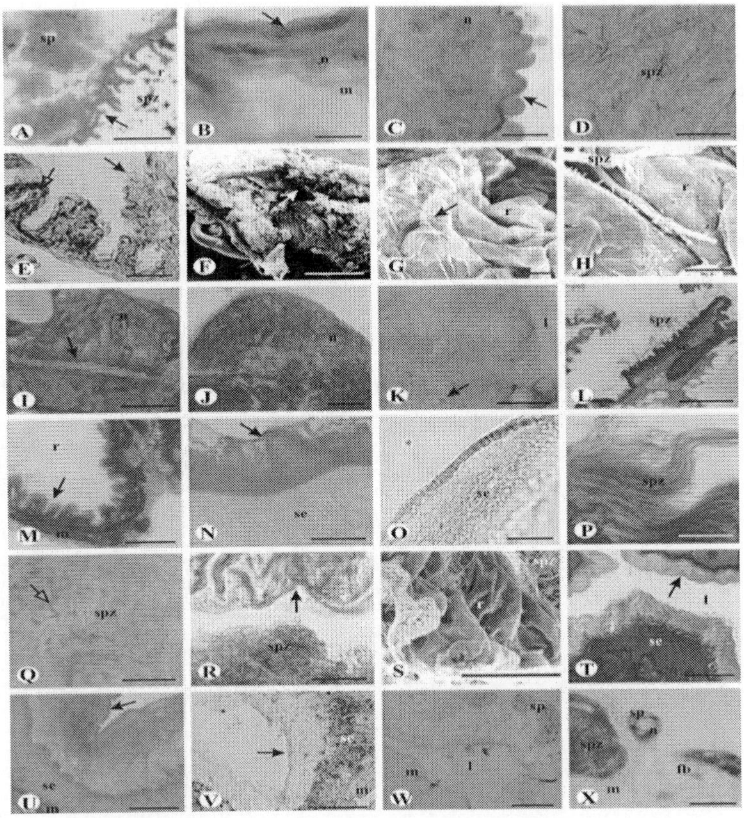

Figure.ureure. 2 A. General view of the spermatheca of A. laevigata stained with Ninhidrina Schiff, Scale bar = 100 μm. B. Detail of the reservoir wall with simple epithelium of scamous cells found in the spermatheca of A. laevigata stained with Ninhidrina Schiff, Scale bar = 20 μm. C. Portion of the stratified epithelium with cubical or polygonal cells found in the spermatheca of A. laevigata stained with Ninhidrina Schiff, Scale bar = 20 μm. D. Spermatozoa in the reservoir of the spermatheca of A. laevigata stained with Bromophenol Blue, Scale bar = 10 μm. E. Criosection of the spermatheca of A. laevigata stained with Sudan Black B, filled block arrow = cuticle, empty block arrow = lipid accumulation at the cells of the coating epithelium, Scale bar = 20 μm. F. SEM of the spermatheca of A. s. sexdens, arrow = interior of the reservoir, Scale bar = 0,5 μm. G. H. SEM of the interior of reservoir of A. s. sexdens, Scale bar G = 0,025 μm, Scale bar H = 8 μm. I. Detail of the spermathecal gland of A. s. sexdens stained with Hematoxyline and Eosin, arrow = canaliculi, Scale bar = 10 μm. J. Detail of the spermathecal gland of A. s. sexdens stained with Ninhidrina Schiff, Scale bar = 10 μm. K. Detail of the spermathecal gland of A. s. sexdens stained with PAS/Alcian Blue, arrow = intracitoplasmic granules PAS positive, Scale bar = 10 μm. L. General view of the spermatheca of A. s. sexdens stained with Hematoxyline and Eosin, Scale bar = 100 μm. M. Detail of the reservoir wall composed by a simple epithelium of scamous cells found in the spermatheca of A. s. sexdens stained with Bromophenol Blue, Scale bar = 50 μm. N. Detail of the internal secretory epithelium of the spermatheca of A. s. sexdens stained with PAS/Alcian Blue, Scale bar = 10 μm. O. Detail of the internal secretory epithelium of the spermatheca of A. s. sexdens stained with Nile Blue, Scale bar = 20 μm. P. Spermatozoa in the reservoir of the spermatheca of A. s. sexdens stained with Bromophenol Blue, Scale bar = 10 μm. Q. Spermatozoa in the reservoir of the spermatheca of A. s. sexdens stained with PAS/Alcian Blue, empty block arrow = secretion with acid polysaccharide, Scale bar = 10 μm. R. Criosection of the reservoir wall of spermatheca of A. s. sexdens stained with Sudan Black B, Scale bar = 10 μm. S. SEM of the interior of reservoir of A. s. piriventris, Scale bar = 0,025 μm. T. General view of the spermatheca of A. s. piriventris stained with Bromophenol Blue, Scale bar = 50 μm. U. Detail of the internal secretory epithelium of the spermatheca of A. s. piriventris stained with PAS/Alcian Blue, Scale bar = 50 μm. V. Criosection of the spermatheca of A. s. piriventris stained with Nile Blue, Scale bar = 20 μm. W. General view of the spermatheca of A. coronatus stained with Hematoxyline and Eosin, Scale bar = 100 μm. X. General view of the spermatheca of C. muelleri stained with Bromophenol Blue, Scale bar = 50 μm.sp = secretory portion, m = musculature, se = internal secretory epithelium, n = nucleus of the epithelial cell, filled block arrow = cuticle, r = reservoir, l = lumen, spz = spermatozoa, fb = fat body cells.

RESULTS AND DISCUSSION

The morphological results obtained for the spermatheca of ant queens of A. laevigata, A. s. sexdens, A. s. piriventris, A. coronatus and C. muelleri revealed that in all five species, this structure is located at beginning of the common oviduct of the female reproductive tract and resembles a bilobed pouch partially jointed at the central portion, therefore exhibiting a common lumen (Figure.ureure. 1). In A. coronatus and C. muelleri, the spermatheca is flatter in shape than those observed in other species (Figure.ureure. 1).

Thus, the spermatheca of queens of A. laevigata, A. s. sexdens, A. s. piriventris, A. coronatus and C. muelleri exhibit a large spermathecal reservoir to store and maintain spermatozoa, in which they remain from mating until egg fertilization (Figure.ureure. 1).

The surface of the reservoir of the spermatheca of A. s. sexdens and A. s. piriventris is rough, and tracheoles are attached into the structure. SEM observations of the internal portion of the reservoir of the spermatheca of these species revealed that this region is internally infolded and irregular, due to the presence of a thick cuticle lining the entire structure (Figure.ureures. 2 F, G, H, S).

Inside the reservoir of A. laevigata, A. s. sexdens, A. s. piriventris and C. muelleri examined in this study, we observed filamentous and elongated spermatozoa with a constant diameter (Figure..ureures. 2 H, S).

To better compare and visualize, histological and histochemical results are summarized in table below:

Aspects of the internal anatomy of ants still remain mostly unknown, as most studies on these insects are focused on control aspects, especially in ants of the genus Atta, considered pests. Although of considerable importance, few studies have been conducted to describe the morphology of the spermatheca of these insects.

In general, the spermatheca of ants has been considered an important structure for their adaptive success, as it receives and stores sperm from mating until egg fertilization, in addition to being a source of origin of new individuals of a colony.

The spermatheca is an organ usually described as a pouch, to which glands might be associated with and discharge their secretion into. These glands are termed spermathecal glands and play a role in the production of secretions that allow the maintenance of a microhabitat with an adequate pH, ionic composition, in addition to other necessary conditions for the survival of spermatozoa.

Most insects present only one spermatheca, as also observed in the queens of the species examined in this study: Atta laevigata, Atta sexdens sexdens, Atta sexdens piriventris, Acromyrmex coronatus and Cyphomyrmex muelleri.

In these species, were confirmed that the spermatheca exhibit a similar morphology resembling a storage pouch described for insects in general, however, with two lobes partially jointed at the median portion of this structure. The same morphology was observed for the spermatheca of queens of the ants A. s. rubropilosa, A. balzani, A. landolti and A. l. balzani (Ortiz; Camargo-Mathias, 2007). In A. coronatus and C. muelleri, the pouch is flatter that those of other species.

Histology and histochemistry of the spermatheca of queens of the ants Atta laevigata, Atta sexdens sexdens, Atta sexdens piriventris, Acromyrmex coronatrus and Cyphomyrmex muelleri

Species Structure	*Atta laevigata*	Atta sexdens *sexdens*	*Atta sexdens piriventris*	*Acromyrmex coronatus*	*Cyphomyrmex muelleri*
Spermathecal gland	- Present as groups of cells forming acini (Figure..ureures. 1, 2 A) - Round and/or polygonal-shaped secretory cells located at the joining point of spermathecal lobes (Figure..ureures. 1, 2 A) - Round-shaped nuclei (Figure..ureures. 1, 2 A) - Ninhydrin Schiff +++ (Figure..ureure. 2 A) - Sudan Black B +	- Present as groups of cells forming acini (Figure..ureures. 1, 2 I, J, K) - Secretory cells varying from cubic to prismatic shaped; cells may be rounder when located near the periphery (Figure..ureures. 1, 2 I, J, K) - Large nuclei consistent with the cell morphology and with 2 to 3 nucleoli (Figure..ureures. 1, 2 I, J, K) - Presence of extracelular cananiculli of considerable caliber internally lined with a thick cuticle (Figure..ureure. 2 I) - Ninhydrin Schiff +++ (Figure..ureure. 2 J) - PAS/Alcian Blue + (Figure..ureure. 2 K) - Sudan Black B –	- Absent	- Present as filaments with a visible central lumen (Figure..ureures. 1, 2 W) - Round to prismatic shaped secretory cells (Figure..ureures. 1, 2 W) - Spherical nuclei located in the median region of cells with evident nucleoli (Figure..ureures. 1, 2 W)	- Present as filaments with a visible central lumen containing secretion (Figure..ureures. 1, 2 X) - Cubic shaped secretory cells (Figure..ureures. 1, 2 X) - Nuclei consistent with the cell morphology and with only 1 nucleolus (Figure..ureures. 1, 2 X) - Bromophenol Blue +++ (Figure..ureure. 2 X)
Internal secretory epithelium	-Absent	- Simple secretory epithelium consisted of prismatic cells and very elongated nuclei located only joining point of the spermathecal lobes (Figure..ureures. 1, 2 L, M, N, O) - Bromophenol Bluel +++ (Figure..ureure. 2 M) - PAS/Alcian Blue – (Figure..ureure. 2 N) - Nile Blue + (Figure..ureure. 2 O)	- Simple secretory epithelium consisted of tall and narrow prismatic cells and nuclei consistent with this morphology. This epithelium, however, differentiates at the joining point of the lobes and cells become even larger (Figure..ureures. 1, 2 T, U, V)	- Absent	- Absent

(Continued)

Species Structure	*Atta laevigata*	Atta sexdens *sexdens*	*Atta sexdens piriventris*	*Acromyrmex coronatus*	*Cyphomyrmex muelleri*
			- Bromophenol Blue +++ (Figure..ureure. 2 T) - PAS/Alcian Blue – (Figure..ureure. 2 U) - Nile Blue +++ (Figure..ureure. 2 V)		
Reservoir wall	- Thin - Consisted of a simple epithelium varying from squamous to cubic (Figure..ureures. 1, 2 B) - Ninhydrin Schiff +++ (Figure..ureures. 2 A, B) - Sudan Black B + (Figure..ureure. 2 E) - At the joining point of spermathecal lobes, this epithelium exhibits several layers of oval or polygonal cells (Figure..ureure. 2 C) - Ninhydrin Schiff +++ (Figure..ureure. 2 C) Sudan Black B –	- Thin, except at the joining point of spermathecal lobes, due to the presence of the prismatic secretory epithelium (Figure..ureures. 1, 2 L, N, O) - Consisted of a simple lining epithelium varying from cubic to squamous (Figure..ureures. 1, 2 M) - Bromophenol Blue +++ (Figure..ureure. 2 M) - PAS/Alcian Blue – (Figure..ureure. 2 N)	- Thick - Consisted of a simple prismatic secretory epithelium (Figure..ureures. 1, 2 T, U, V) - Bromophenol Blue +++ (Figure..ureure. 2 T) - PAS/Alcian Blue – (Figure..ureure. 2 U) - Nile Blue +++ (Figure..ureure. 2 V)	- Thin - Consisted of a simple squamous epithelium exhibiting cells with flattened nuclei (Figure..ureures. 1, 2 W)	- Thin - Consisted of a simple squamous epithelium (Figure..ureures. 1, 2 X) - Bromophenol Blue + (Figure..ureure. 2 X)
Cuticle found internally lining the spermatheca	- Thick in most of its area (Figure..ureures. 1, 2 A, B) - Ninhydrin Schiff +++ (Figure..ureures. 2 A, B) - Sudan Black B + (Figure..ureure. 2 E) - Thin on the stratified epithelium (Figure..ureure. 2 C)	- Thick throughout (Figure..ureures. 1, 2 M, N, R) - Bromophenol Blue +++ (Figure..ureure. 2 M) - PAS/Alcian Blue ++ (Figure..ureure. 2 N) - Sudan Black B + (Figure..ureure. 2 R)	- Thick, but it becomes thinner in the joining point of the lobes where the epithelium differentiates (Figure..ureures. 1, 2 T, U) - Bromophenol Blue+ (Figure..ureure. 2 T) - PAS/Alcian Blue ++ (Figure..ureure. 2 U) - Nile Blue – (Figure..ureure. 2 V)	- Thin	- Thin (Figure..ureures. 1, 2 X) - Bromophenol Blue + (Figure..ureure. 2 X)

Species Structure	*Atta laevigata*	Atta sexdens *sexdens*	*Atta sexdens piriventris*	*Acromyrmex coronatus*	*Cyphomyrmex muelleri*
Musculature found externally surrounding the spermatheca	- Several layers (Figure..ureures. 1, 2 B) - Ninhydrin Schiff +++ (Figure..ureure. 2 B) - Sudan Black B –	- Several layers (Figure..ureures. 1, 2 M) - Bromophenol Blue +++ (Figure..ureure. 2 M) - PAS/Alcian Blue + - Sudan Black B +	- Several layers (Figure..ureures. 1, 2 U, V) - Bromophenol Blue +++ - PAS/Alcian Blue +++ (Figure..ureure. 2U) - Nile Blue – (Figure..ureure. 2 V)	- Few layers (Figure..ureures. 1, 2 W)	- Few layers (Figure..ureures. 1, 2 X) -Bromophenol Blue + (Figure..ureure. 2 X)
Morphology of the spermatozoa	- Elongated, no distinction between head and tail (Figure..ureures. 1, 2 D) - Bromophenol Blue +++ (Figure..ureure. 2 D)	- Elongated, no distinction between head and tail (Figure..ureures. 1, 2 P, Q, R) - Bromophenol Blue +++ (Figure..ureure. 2 P)- PAS/Alcian Blue ++ (Figure..ureure. 2 Q) - Sudan Black B ++ (Figure..ureure. 2 R)	- Elongated, no distinction between head and tail (Figure..ureures. 1, 2 S)	- Elongated, no distinction between head and tail (Figure.ureure. 1)	- Elongated, no distinction between head and tail (Figure..ureures. 1, 2 X) - Bromophenol Blue +++ (Figure..ureure. 2 X)

+ Weak positive reaction.
++ Medium positive reaction.
+++ Strong positive reaction.
Negative reaction.

The presence of a fat body around organs has been reported for most insects, as observed surrounding the spermatheca of C. muelleri, however, it is unclear whether the fat body has any influence on the spermatheca or it only plays a role filling the abdomen. In the Orthoptera Locusta migratória, the spermatheca is also surrounded by a layer of fat body, probably acting as a protective structure in the pupa as well as in the adult (Lay and Hartmann, 1999).

Another characteristic aspect of the spermatheca of queens of most insects is the presence of tracheae. According (Poole, 1970), these tracheae are responsible for gas exchanges, which are of extreme importance to maintain the integrity of epithelial cells and consequently fully functional.

The secretory cells of the spermatheca of most insects exhibit the membrane with abundant microvilli, which increase the cell surface for exchanges, indicating that the secretory epithelium may play a role in the reabsorption of the contents present in the lumen, in addition to cuticular support. Apical microvilli in the epithelial cells of the spermatheca have been described in some species of Blattaria (Gupta and Smith, 1969), Coleoptera (Happ and Happ, 1975; Tombes and Roppel, 1972), Orthoptera (Lay and Hartmann, 1999), and in the stingless bee Melipona bicolor (Cruz-Landim and Serrão, 2002). In the present study, similar structures, resemble a brush border, were observed in the internal secretory epithelium of A. s. piriventris. In the spermatheca of these study species, as well as species of genus Atta and Acromyrmex examined by Ortiz and Camargo-Mathias (2007), regions probably playing a secretory role exhibited two morphologies: a) groups of cells forming structures resembling acini or filaments, located externally to the structure (spermathecal gland), and b) as a secretory epithelium internally lining the reservoir. Thus, depending on the species, the spermathecal secretion may be produced in three sites: a) in the spermathecal gland located in the external portion, b) in the secretory epithelium located in the internal portion of the reservoir or c) in both portions.

A spermathecal gland was observed in queens of A. laevigata, A. s. sexdens and A. coronatus, and characterized by groups of cells, and in C. muelleri, by filaments. The presence of this gland forming groups of cells resembling acini was also observed by Ortiz and Camargo-Mathias (personal communication) in the species A. s. rubropilosa and A. l. balzani. In the spermatheca of queens of A. s. piriventris, however, the external associated gland was absent, as also reported by Ortiz and Camargo-Mathias (2007) in the species A. balzani and A. landolti, suggesting a morphological similarity among the spermatheca of these ant species.

The spermatheca of the species A. laevigata, A. coronatus and C. muelleri present a reservoir with thin walls, due to the absence of an internal secretory epithelium, which was only present in A. s. sexdens, located only at the joining point of the spermathecal lobes and in A. s. piriventris, internally lining the entire spermatheca. The presence of an internal secretory epithelium in the spermatheca of queens of A. s. piriventris and A. s. sexdens suggests that the physiology to maintain the sperm in viable conditions for egg fertilization might differ from those of A. laevigata, A. coronatus and C. muelleri, in which only the external spermathecal gland was observed.

Differences in thickness epithelium of the spermatheca are correlated with its roles, whether secretory or storing. In general, the epithelium of the spermatheca of insects is classified in three basic types, according to the location of secretory cells (Gillot, 1988). In the first type, found in Periplaneta americana, secretory cells are distributed throughout the entire wall of the reservoir (Gupta and Smith, 1969), as also observed in our study for A. s.

piriventris, and the ant species A. l. balzani, A. balzani and A. landolti examined by Ortiz and Camargo-Mathias (2007). In the second type, gland cells are restricted to a particular region of the epithelium, as reported for Aedes aegypty (Clements and Potter, 1967), and for A. s. sexdens examined in our study. In the third type, gland cells form a gland distinct from the reservoir, as found in most Hymenoptera and several Coleoptera (Dallai, 1975, Happ and Happ, 1975), in queens of A. laevigata, A. s. sexdens, A. coronatus and C. muelleri examined in present study, and in A. s. rubropilosa described by Ortiz and Camargo-Mathias (2007).

According to the observed in the queens of A. laevigata, A. s. sexdens, A. coronatus and C. muelleri, were divided the spermatheca of these ant species in two distinct portions: a) a secretory portion represented by the spermathecal gland and consisted of cells responsible for the production the spermathecal liquid and b) a storing portion, represented by the reservoir consisted of a simple lining epithelium, where spermatozoa are stored and nourished until egg fertilization. Specifically in queens of A. s. sexdens, the secretory portion of the spermatheca was subdivided once more in: a) external secretory portion, represented by the spermathecal gland located externally to the structure and b) internal secretory portion, represented by the secretory epithelium located internally to the reservoir and restricted to the joining point of spermathecal lobes. Since in A. s. piriventris the spermatheca is not divided into two distinct portions, and the epithelium producing the secretion is located internally to the reservoir, the secretion is produced and discharged directly by exocytosis, with vesicles from the apical portion of the cytoplasm of secretory cells being released into the lumen of the spermatheca, without the participation of secretory cells externally to the organ. The same process has already been observed and described for the spermatheca of Rhodnius prolixus (Gillot, 2003).

In A. laevigata, A. s. sexdens, A. coronatus, and C. muelleri, the external secretory cells form a central collecting canaliculus, in which intracellular canaliculi open and collect the secretion produced by these cells, as well as transport them to the lumen of the reservoir. When we examined queens of A. s. sexdens, which exhibit two types of secretory epithelium, it is worth pointing out that canaliculi were only observed in cells of the spermathecal glands, while they were absent in the internal secretory epithelium. This indicates a different way, by exocytosis, to release the secretion produced by the cells of epithelia located internally to the reservoir.

Histochemically, these results revealed that in the spermatheca of two of three species of the genus Atta and that exhibit an internal secretory epithelium, A. s. sexdens and A. s. piriventris, the epithelium was strongly positive for proteins, indicating that this element is produced in large quantities and probably plays an important role in the composition of the secretion. The spermatheca of C. muelleri was also positive for this histochemical test, specially the spermathecal gland, the spermatozoa, and the secretion present in the lumen of the reservoir, while the lining epithelium and muscles externally surrounding the spermatheca were weakly stained.

Tests demonstrating polysaccharides revealed that secretory cells, internally located in the spermatheca of A. s. sexdens and s. piriventris, were weakly stained or unstained in certain regions. This was also observed in the spermathecal gland of A. s. sexdens and laevigata. In the cytoplasm of these cells, only small PAS-stained deposits resembling fine granules were observed, indicating low levels of production of these elements in the spermatheca of the three species. Gardner (2004) reported that the spermatheca of Schistocerca americana and Dissosteira carolina were weakly to moderately stained by PAS, while the cuticle lining this structure was strongly positive for acidic polysaccharides.

In the spermatheca of A. laevigata, a large amount of polysaccharides was observed in the cytoplasm of cells of the epithelium lining the reservoir, which does not play a secretory role in the production of components for the maintenance of spermatozoa in this species, suggesting that, since the cuticle is a structure that exhibits large quantities of polysaccharides and proteins, these polysaccharides might be produced by the cells of this epithelium.

The low levels of polysaccharides in the spermatheca of the studied species of the genus Atta, as well as those of other insects, suggest that these elements might take part only in the formation of complexes with other compounds, such as glycoproteins, which might play a role in the maintenance of spermatozoa, providing them a s source of energy while stored inside these structures and mainly, providing energy necessary for their metabolism, including during the time near fertilization.

Tests performed in these study demonstrated that lipids were more abundant in the spermatheca of A. s. piriventris, although some staining was observed in the secretory cells of A. s. sexdens and A. laevigata. Weak staining by Sudan Black B was also observed by Gardner (2004) in the spermatheca of S. americana e D. carolina.

These results obtained for lipids showed a low production of these elements in spermathecae, suggesting that, similarly to polysaccharides, they might be produced only to form complexes with other compounds. The small amounts of polysaccharides and lipids found in the spermathecae of ants of the genus Atta, as opposed to the large quantities of proteins, suggest that the spermathecal secretion primarily consists of proteins, but that form complexes with lipids and polysaccharides.

Blum et al. (1967), in some studies on the reproductive system of bees, reported that phospholipids can also used as a source of energy, although in this case, oxidative energy for respiration. The use of this compound might be associated to the fact that, under aerobic conditions, the spermatozoon might be mobile for at least three hours before ejaculation.

There are few studies specifically on the secretion produced by spermathecal secretory cells, as well as the secretion present in the spermatheca. Some of them, however, have demonstrated that the secretion present in the lumen of the spermatheca, product of secretory cells, is of glycoprotein nature (Ahmead and Gillot, 1982a; Ahmead and Gillot, 1982b; Bhatinagar and Musgrave, 1971; Clements and Potter, 1967). The spermatozoa utilize this secretion as a source of energy, and are in constant movement inside this structure.

The spermatozoa of most insects have been described as elongated and filamentous structures (Baccetti, 1972; Dallai and Afzelius, 1993), as also observed in all ant species examined in the present study. Histochemical tests revealed that spermatozoa of these ants species are elongated and uniform structures consisted primarily of polysaccharides and proteins, as lipids were only observed in the membrane of spermatozoa of A. laevigata and A. s. sexdens.

Conclusion

The old saying that prevention is better than cure also serves to cutting control in crops. When it is necessary to make agriculture planning the actions. As a farmer plowing plans, disking, have to plan the control of ants as well. The farmer has to keep in mind you need, first of all, kill ant at the right time, apply the insecticide only correct and then make the ground handling operation. If you put heavy machinery work the land before settling ants, can come some annoyances. With cane sugar, for example, which control method is the most common thermonebulization, there is a difficulty in Figure.ureurehting the ants own mechanization. The heavy machinery and even the disruption of soil neutralized the nests, destroying some of them. If channels of the nest are not perfect, the toxic fumes will not move properly and is unlikely to reach the fungus chambers. Hence the need to plan. If everyone planning, we would not have large problems with ants.

These preliminary morphological datas on the spermatheca of queens of A. s. sexdens, A. s. piriventris, A. laevigata, A. coronatus and C. muelleri revealed that the latter three species present only an external secretory epithelium (spermathecal gland); A. s. sexdens exhibits an external as well as an internal secretory epithelium; and A. s. piriventris presents only an internal secretory epithelium.

Since A. s. piriventris exhibits a secretory epithelium only in the internal region of the spermatheca, this suggests that the spermatheca itself might be self-sufficient to produce all elements of secretion in order to maintain the spermatozoa viable, without the need of exogenous elements from the hemolymph. The little or no dependence on exogenous compounds in relation to the spermatheca that contribute to provide a suitable microhabitat for the spermatozoa until egg fertilization might increase the chances of survival of the species, even during unfavorable environmental conditions.

REFERENCES

Ahmead, L; Gillot, C. (1982a). The spermathecae of *Melanoplus sanguinipes* (Fabr.) I. Morphology, histology and histochemistry. *Int. J. Invert. Rep*, 4, 281-295.

Ahmead, L; Gillot, C. (1982b). The spermathecae of *Melanoplus sanguinipes* (Fabr.) II. Ultrastructure. *Int. J. Invert. Rep*, 4, 297-309.

Baccetti, B. (1972). Insects sperm cell. *Advances in Insects Physiology*, London., 9, 315-395.

Ballari, SA; Farji-Brener, AG. (2006). Refuse dumps of the leaf-cutting ants as a deterrent for ant herbivory: does refuse age matter?.*The Netherlands Entomological Society,* 121, 215–219.

Berry, DJ. (1985). Material direction of oogenesis and early embryogenesis in insects. *Annu. Rev. Entomol*, 27, 205-227.

Bhatinagar, RD; Musgrave, AJ. (1971). Aspects of the histophysiology of the spermathecal gland of Stophilus granarius (L) (Coleóptera). *Canad. J. Zool*, 49, 275-277.

Blum, MS; Glowska, Z; Taber, S. (1967). Composition and possible significance of fatty acids in the lipid classes in honey bee semen. *Journal of Insect Physiology*, 13, 1301-1308.

Bot, ANM; Currie, CR; Hart, AG; Boomsma, JJ. (2001). Waste Management in Leaf-cutting Ants. *Ethology Ecology and Evolution,* 13, 225–237.

Chapman, RF. (1998). *The insects: structure and function*. New York, Elsevier, pp.770.

Clements, AN; Potter, SA. (1967). The fine structure of the spermathecae and their ducts in the mosquito *Aedes aegypti*. *J. Insect. Physiol*, 13, 1825-1836.

Cruz-Landim, C; Serrão, JE. (2002). Ultrastructure of the spermathecal gland of Melipona bicolor Lep. (Hymenoptera, Apinae, Meliponini). *Braz. J. Morphol. Sci*, 1, 9-16.

Currie, CR; Mueller, UG; Malloch, D. (1999). The agricultural pathology*PNAS*, 96, 7998–8002.

Dallai, R. (1975). Fine structure of the spermathecae of *Apis Mellifera*. *J. Insect Physiology*, 21, 89-109.

Dallai, R; Afzelius, B. (1993). Axonemal structure and insect phylogeny. *Boll. Zool*, 60, 423-429.

Davey, KG. (1985). The female reproductive tract. In: Kerkut GA, Gilbert LI (Ed). Comprehensive Insect Physiology: Embryogenesis and Reproduction. Pergamon Press.

Gardner, GE. (2004). Morphological and histological aspects of the spermatheca as they relate to sperm organization in the grasshopper species Schistocerca americana and Dissosteira carolina (Orthoptera: Acrididae). 2004, 41f. Thesis (Degree of Master of Science) - Faculty of North Carolina State University, Raleigh.

Gillot, C. (1988). *Insecta. In Reproductive Biology of Invertebrates*. Vol. III. Acessory Sex glands. Edited by K.G. Adiyodi and R.D. Adiyodi. John Wiley and sons Ltd., U.K., p. 356-471.

Gillot, C. (2003). Male acessory gland secretions: Modulations of female reproductive physiology and behaviour. *Annu. Rev. Entomol*, 48, 163-184.

Gupta, BL; Smith, DS. (1969). Fine structural organization of the spermatheca in the cockroach *Periplaneta americana.Tissue and Cell*, 1, 295-324.

Happ, GM; Happ, CM. (1975). Fine structure of the spermathecae of the mealworm beetle *Tenebrio molitor* L. *Cell Tissue Res*, 162, 253-269.

Hart, AG; Ratnieks, FLW. (2002), Waste management in the leaf-cutting ant *Atta colombica. Behavioural Ecology*, 13, 224–231.

Hölldobler, B; Wilson, EO. (1990). *The ants*. Cambridge: The Belknap Press of Harvard University, pp. 732.

Lay, M; Zissler, D; Hartmann, R. (1999). Ultrastructural and functional aspects of the spermatheca of the African Migratory Locust Locusta migratoria migratorioides (Reiche and Fairmaire) (Orthoptera: Acrididae). *Int. J. Insect Morphol. Embryol*, 28, 349-361.

Ortiz, G; Camargo-Mathias, MI. (2007). Spermatheca of four species of ants of the tribe Attini (Hymenoptera: Myrmicinae). Morphological specialization. *Research Journal of Biological Sciences*, 2, 5-12.

Piper, R. (2007). Extraordinary Animals*: An Encyclopedia of Curious and Unusual Animals*, Greenwood Press, pp. 298.

Poole, HK. (1970). The wall structure of the honey bee spermatheca with comments about its function. *Ann. Entomol. Soc. Am*, 63, 1625-1628.

Reynolds, HT; Currie, CR. (2004). Pathogenicity of Escovopsis weberi: The parasite. *Mycologia*, 96: 955–959.

Speight, MR; Watt, AD; Hunter, MD. (1999). Ecology of Insects. *Blackwell Science*. pp.156.

Tombes, AS; Roppel, RM. (1972). Ultrastructure of the spermatheca of the granary weevil Sitophilus granarius (Coleoptera, Curculionidae). *Int. J. Insect. Morphol. Embryol*, 1, 141-152.

Thornhill, R; Alcock, J. (1983). *The evolution of insect mating systems*. Harvard Press, Cambridge, Mass.

Weber, NA. (1972). Gardening Ants, The Attines. *The American Philosophical Society*, p.14-34.

Zhang, MM; Poulsen, M; Currie, CR. (2007). Symbiont recognition of mutualistic bacteria by *Acromyrmex* leaf-cutting ants. *The ISME Journal*, 1, 313–320.

In: Invasive Species
Editors: Joaquín J. Blanco and Adrian T. Fernandes
ISBN: 978-1-61942-761-7

Chapter 6

FRESHWATER POLLUTION AND BIOLOGICAL WATER QUALITY MONITORING: THE INCREASING PROBLEM OF INVASIVE SPECIES AND ACCURATE ECOLOGICAL ASSESSMENT*

Calum MacNeil†
The Government Laboratory, Ballakermeen Road,
Douglas, Isle of Man IM1 4BR, British Isles

ABSTRACT

Macroinvertebrates are routinely used in the calculation of indices to assess the quality of rivers throughout the world. Macroinvertebrate assemblages have proved themselves to be reliable, cost effective indicators of freshwater pollution, as the relative sensitivities of their members respond to changes in water quality in predictable ways. Such biological monitoring can have advantages over chemical monitoring, as it tends to be cheaper and more reliable over the long term, especially in rivers where water quality can fluctuate dramatically on a day to day basis. However, an increasing threat to accurate biological monitoring comes from invasive species. The rate of freshwater invasions is increasing, with increasing global transport, climate change and environmental degradation all facilitating the arrival of pollution tolerant and often aggressive invaders in many previously undisturbed river systems. Such invasive species can have a major impact on benthic macroinvertebrate assemblage structure and diversity. Macroinvertebrate invaders often radically alter native macroinvertebrate assemblages through biotic interactions, such as competition and predation. In many cases more pollution tolerant invaders can replace more sensitive natives and very predatory invaders can decimate native assemblages as a whole. This has serious implications for accurate biological monitoring and the consequent assessment of the ecological status of invaded rivers. The whole structure of existing water quality indices

* A version of this chapter also appears in Pollution Monitoring, edited by Alicia C. Ortiz and Nancy B. Griffin, published by Nova Science Publishers, Inc. It was submitted for appropriate modifications in an effort to encourage wider dissemination of research.

† Email:calummanx@hotmail.com; calum.macneil@lab.dlge.gov.im.

may need revising to take into account the presence of invasive species within monitored assemblages, such that the accuracy of any derived assessments of water quality can be judged in the context of whether the assemblages generating them contain mainly natives or invaders. In addition, methods of 'biocontamination' assessment, designed to be used alongside current routine water quality indices may represent a simple, cost-effective way to integrate the biological pressure invasives represent, into a country's or region's routine river quality monitoring programme. Such methods should be adopted alongside routine biological monitoring of pollution to provide a more comprehensive estimate of ecological quality status. It is clear that he problems of invasive species and accurate monitoring of freshwater pollution can no longer be ignored.

INTRODUCTION

Biological invasions pose an increasing threat to biodiversity [Sala *et al.* 2000; Clavero and García-Berthou, 2005; Ding *et al.*, 2008; Strayer, 2010], with climate change increasing the impacts of invaders [Tyliankis *et al.*, 2008] and increased global transport accelerating the rate of invasions, particularly in freshwaters [Kinzelbach, 1995; Strayer, 2010]. Benthic macroinvertebrates feature in many such invasions [Conlan, 1994; MacNeil *et al.*, 1999a, 2009; Crawford *et al.*, 2006; Leuven *et al.*, 2009] and the ecological impacts of such introductions can be profound [Pinkster *et al.*, 1992; Ricciardi, 2001; van Riel *et al.*, 2006]. For instance, many of these invasive species can irreversibly change native assemblage structure through competition and predation, with native species eliminated and replaced [Ricciardi *et al.*, 1998; Kelly *et al.*, 2002, 2003, 2006; Crawford *et al.*, 2006].

Freshwater invasions frequently accompany pollution or environmental degradation or follow major disturbances [den Hartog *et al.*, 1989; Conlan, 1994; MacNeil *et al.*, 2004]. The EU Water Framework Directive [WFD] represents an important tool designed to facilitate improvements in water quality across Europe [European Commission, 2000]. The WFD requires the maintenance of high ecological status in freshwaters where it already exists [i.e. near pristine macroinvertebrate assemblages] and achieving a minimum of good ecological status in all freshwaters by 2015. Implementation of ecological assessments for the WFD has prompted development of assessment schemes based on biological elements geared to detect measurable responses to specific pressures [Solheim and Gulati, 2008]. Benthic macroinvertebrate assemblages via the calculation of biotic indices [Armitage *et al.*, 1993; Rosenberg and Resh, 1993; Wright *et al.*, 1993] represent an important biological element routinely used to assess river water quality for ecological classification for WFD purposes [MacNeil, 2006]. Unfortunately, invasive species are a major biological pressure and threat to this core objective of the WFD [Orendt *et al.*, 2009], with many European river systems increasingly dominated by ecologically damaging invaders, including crayfish, amphipods and bivalves [Arbačiauskas *et al.*, 2008; Cardoso and Free, 2008; MacNeil and Briffa, 2009; Panov *et al.*, 2009].

Biotic indices such as the Biological Monitoring Working Party (BMWP) score and the Average Score Per Taxon (ASPT) [Biological Monitoring Working Party, 1978; Armitage *et al.*, 1983] are routinely used in the ecological status assessment of rivers in the U.K. The BMWP system is similar to many other biotic indices, in that it assigns scores to macroinvertebrate families based on their perceived relative sensitivities or tolerances to organic enrichment. However, some invasive species are capable of surviving rigorous introduction mechanisms that allow them to enter river systems where environmental degradation has diminished native assemblages [den Hartog *et al.*, 1989; Conlan, 1994; MacNeil *et al.*, 2004] and are far more pollution tolerant than the natives they may be replacing [Dick and Platvoet, 1996; MacNeil *et al.*, 2000]. A major problem that results when such invaders establish themselves within native assemblages, is that the BMWP and other similar indices do not distinguish between native and invasive species within the same family, even when they differ in sensitivity to organic pollution [Walley and Hawkes, 1996; MacNeil *et al.*, 2000] Given the current European situation with many rivers being increasingly degraded and polluted [Andrew, 1998; Panov *et al.*, 2009], species replacements are likely to increase. Therefore the potential for the function and accuracy of biotic indices to be compromised by a native species being substituted by an invader requires urgent investigation [Orendt *et al.*, 2009].

The Problem of Biotic Indices and Accurate Ecological Assessment – When the Invader Is More Pollution Tolerant Than the Native

The amphipod crustacean family Gammaridae, is an example of BMWP scoring family which contains many native and invasive species with widely differing physiological tolerances [Walley and Hawkes, 1996; Gaston and Spicer, 2001]. *Gammarus* spp. have featured in many invasions because of both accidental and deliberate introductions linked to aquaculture, angling and ecological experiment [Hynes, 1950, 1954; MacNeil *et al.*, 1999a]. In a computer based reappraisal of BMWP scores in England and Wales, the presence of the invasive North American *Gammarus tigrinus* which may be more pollution tolerant than native *Gammarus* spp., was proposed as a reason for a large disparity in the current BMWP score assigned to this family and the lower score that the reappraisal suggested [Walley and Hawkes, 1996]. Indeed, when the BMWP system was introduced, it was acknowledged that "the score system will probably need to be modified in the light of practical experience" [Biological Monitoring Working Party, 1978].

An example of a modification required is that of the well studied replacement of a native amphipod *Gammrus duebeni celticus* by an invader *Gammarus pulex* [reviewed in Dick, 2008]. Since the 1940s and 50s, populations of *G. d. celticus* in the Isle of Man [an island of 500km^2 in the Irish Sea, 26km from mainland Britain] and Northern Ireland respectively, have been subjected to invasion by *G. pulex* deliberately introduced from mainland Britain [Strange and Glass, 1979; Dick *et al.*, 1997; MacNeil *et al.*, 2009]. *G. d. celticus* and *G. pulex* are 'lumped' together as 'Gammaridae' in the BMWP index and assigned the same score [MacNeil *et al.*, 2000]. In many river systems in N. Ireland and the Isle of Man, *G. d. celticus* has replaced *G. pulex* [Dick, 2008], through intraguild predation or IGP (predation between competitors belonging to the same ecological guild - see Polis *et al.*, 1989), with lower dissolved oxygen levels and higher levels of organic pollution favouring IGP of *G. d. celticus* by *G. pulex* [MacNeil *et al.*, 2004]. Large differences in the sensitivity of the two species could result in erroneous scores in the BMWP scoring system and this was explicitly tested by MacNeil and Briffa [2009] using an extensive data-set (119 sites) from N. Ireland (61 sites in Lough Neagh and the Lagan catchments) and the Isle of Man (58 sites), with water quality assessed by the ASPT derivation of the BMWP. The ASPT is the BMWP score divided by the number of families that generated it and is regarded as a stable and reliable index of organic pollution [Armitage *et al.*, 1983]. N. Ireland has fewer macroinvertebrate taxa than mainland Britain and the Isle of Man has only 87% of the BMWP scoring families of mainland Britain [Styles, 2001] and therefore the ASPT index was regarded as more reliable for comparisons between countries than the purely additive BMWP. MacNeil and Briffa [2009] generated two ASPT values for each site, one with the Gammaridae included and one with this group removed. When the biotic index was calculated for native and invader sites [*Gammarus* spp. included], it became clear that the invader occurred with macroinvertebrate assemblages more tolerant of organic pollution, while the native occurred with more sensitive assemblages, as indicated by the BMWP system [MacNeil and Briffa, 2009] [Figure 1].

When both *Gammarus* spp. were excluded from calculation of the indices, it was also clear that the presence of *G. pulex* falsely elevated the biotic index score, while the presence of the native *G. d. celticus* did not [Figure 2]. This 'over-inflation' of scores would be most evident in the poorest quality areas, where *G. pulex* would constitute the highest scoring BMWP group, while co-occurring with very tolerant taxa such as isopods, chironomids and oligochaetes. These findings implied that *G. pulex*, at least in the parts of its range where it is considered as an invader, should be given lower scores relative to *G. d. celticus*, in terms of pollution tolerance and sensitivity in the BMWP, ASPT and similar indices.

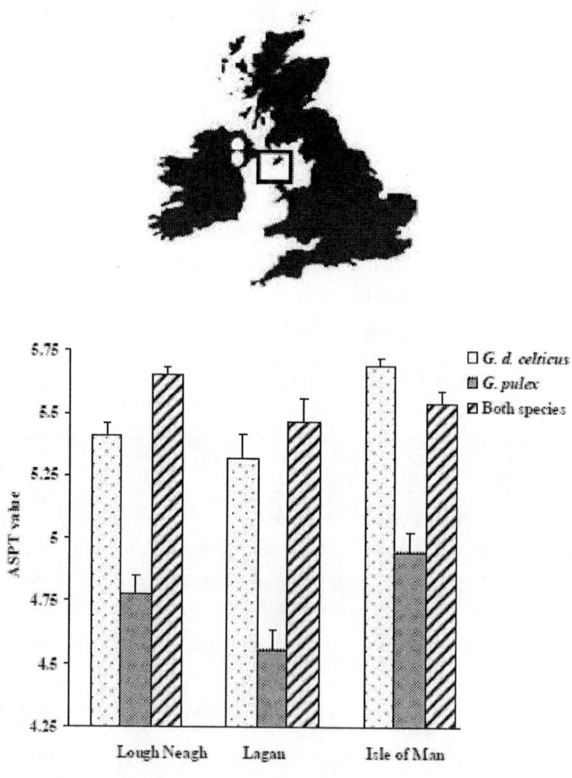

Figure 1. Mean + SE ASPT value of river sites containing either native *G. d. celticus* or invasive *G. pulex* in Lough Neagh, Lagan and Isle of Man river networks (inset map shows locations of L. Neagh (top circle) and Lagan (bottom circle) river systems in N. Ireland and the Isle of Man (square)) (MacNeil *et al.*, 2010).

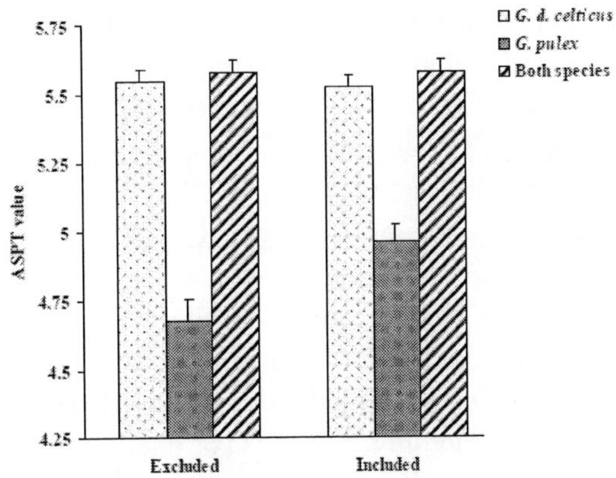

Figure 2. Mean + SE ASPT value of river sites containing either native *G. d. celticus* or invasive *G. pulex*, with amphipods both excluded from and included in ASPT index generation (MacNeil and Briffa, 2009).

Walley and Hawkes [1996] suggested that the BMWP score for Gammaridae should be downgraded from 6 to 4 [especially when compared to the mayfly family Baetidae which only scored 4] because of the presence of tolerant invasives. This view is further reinforced when the North American *Crangonyx pseudogracilis* Bousfield is considered, as this is found in even lower water quality areas than *G. pulex* in N. Ireland and the Isle of Man [MacNeil *et al.*, 2000, 2004]. For instance, although *G. pulex* has replaced *G. d. celticus* in lower water quality areas of rivers, in even more grossly polluted areas, both these *Gammarus* spp. are absent and only *C. pseudogracilis* is present. Unfortunately, the family Crangonyctidae is grouped with the Gammaridae in the BMWP scoring system, increasing the scope for flawed assessments of biological water quality [MacNeil *et al.*, 2000].

Again, the influence of relatively high scoring amphipods such as *C. pseudogracilis* will be disproportionately high in very poor water quality sites where there tends to be a depauperate and low scoring assemblage and this disparity will increase as water quality decreases.

The Problem of Biotic Indices and Accurate Ecological Assessment – When the Invader Is More Predatory Than the Native

Another major problem with invasive species occurs when the invader may be more predatory of other members of the native assemblage. This has obvious implications for river water quality monitoring as the macroinvertebrate taxa which contribute to biotic indices such may be differentially impacted upon by the native and invader. Returning to the replacement of *G. d. celticus* by *G. pulex* in both N. Ireland and the Isle of Man, there are very few sites with both *G. d. celticus* and *G. pulex* co-occurring, even though the majority of rivers contained both species. This may indicate that once *G. pulex* has invaded a site, it is difficult for the two species to coexist long-term and Hynes [1954, 1955] thought that *G. duebeni* is invariably eventually replaced on any land mass invaded by *G. pulex*. Regardless of water quality, invasive *G. pulex* may also be more predatory than the native *G. d. celticus* to a wide range of co-occurring macroinvertebrates [MacNeil *et al.*, 1997; Kelly *et al.*, 2002, 2003]. For instance, assemblage composition, biomass and diversity differed markedly between *G. pulex* and *G. d. celticus* dominated areas in contiguous reaches of a N. Irish river, where water chemistry remained constant [see Kelly *et al.*, 2003, 2006]. Kelly [2006] ascribed these differences to increased competition and predation by *G. pulex*.

Another invader which may cause even greater disruption to water quality monitoring is the Ponto-Caspian *Dikerogammarus villosus*, which has invaded much of Central Europe and has very recently arrived in the British Isles [MacNeil *et al.*, 2010] This amphipod may become super-abundant within invaded sites [van Riel *et al.*, 2006] and is highly predatory not only of other native and invasive amphipods (including *G. duebeni*, *G. pulex* and *G. tigrinus*) but of a wide range of macroinvertebrate taxa, including mayfly, isopods, damselfly and chironomids which feature in indices [Dick *et al.*, 2002; MacNeil and Platvoet, 2005]. The potential impacts of this particular amphipod on water quality monitoring will probably be profound, but have not yet been examined. Gabriels *et al.* [2005)] did however note that the *D. villosus* may outcompete a number of native amphipod species, but even then, this might not influence the results of a biotic index (Belgian Biotic Index or BBI) calculation at family level of a sample as Gammaridae may still be present.

Biocontamination – A Way to Integrate the Presence of Invaders into Routine Water Quality Monitoring?

Cardoso and Free [2008] point out there is increased debate on how to incorporate invaders into ecological assessments owing to their acknowledged potential to significantly alter the structure and functioning of aquatic ecosystems. Although invaders constitute a major biological pressure to objectives of the WFD [Arbačiauskas *et al.*, 2008], the current approaches to ecological status or water quality assessment invariably ignore the presence of invaders. This 'willful ignorance' can result in a site with a macroinvertebrate assemblage containing a high proportion of invaders still being classed as having good ecological condition or good water quality [Arbačiauskas *et al.*, 2008; Cardoso and Free, 2008; MacNeil and Briffa, 2009]. It does seem clear that the BMWP score and related/similar indices needs to take into account the presence of invasives, whether they be amphipods [MacNeil *et al.*, 2000; Dick *et al.*, 2002], crayfish [i.e. signal crayfish *Pacifastacus leniusculus* - Crawford *et al.*, 2006] or bivalves [i.e. zebra mussel *Dreissena polymorpha* - Strayer and Malcolm, 2007].

Realising the need to integrate the presence of invaders within established routine monitoring programmes, Arbačiauskas *et al.* [2008] proposed a simple method to assess 'biocontamination', a term which refers to the presence of an invasive species [Arbačiauskas *et al.* 2008], regardless of its ecological impact [see Colautti and MacIsaac, 2004; Ricciardi and Cohen, 2007]. This method has been designed to utilize routine water quality monitoring data and thus, should require minimal extra effort, time, expertise or expense in its generation to that already expended for WFD purposes or similar. Five classes of biocontamination were defined that corresponded to five ecological quality classes designated for WFD purposes [European Commission, 2000] allowing status ranking from 'high' down to 'bad', with the impact of invaders on native macroinvertebrate assemblages assumed to be proportional to the occurrence and abundance of invaders within resident assemblages. This is taken into account by the proposed index and its related metrics [Arbačiauskas *et al.*, 2008)]. Site-specific indices of abundance [the abundance contamination index, ACI] and order/family richness [the richness contamination indices, the RCIo and RCIf] contamination are calculated to generate site-specific biocontamination index [SBCI] classes [no, low, moderate, high or severe] for all sites. For instance, 'bad' ecological status class is designated when invaders constitute over 50% of the orders or 50% of the abundance of the assemblage, as it is assumed that such an assemblage could only have developed as a result of invasion [Arbačiauskas *et al.*, 2008]. Indeed, Arbačiauskas *et al.* [2008] showed that many of the main inland waterways of Europe were highly contaminated with invasives, with abundance contamination exceeding 50% in the Nemunas, Oder, Rhine, Main and Danube rivers.

The original testing of the proposed biocontamination index was conducted on a large, European-wide scale, within rivers and lakes in Central Europe [Arbačiauskas *et al.*, 2008] and was reliant on macroinvertebrate data derived from different operators, using disparate sampling methods and where differing biogeographical factors had to be accounted for and where measures of chemical water quality were unavailable [Arbačiauskas *et al.*, 2008]. However, despite these limitations it was shown to be a useful simple metric and supplement to the BMWP scheme. The testing was advanced further by MacNeil *et al.* [2010] who applied the biocontamination index to the Isle of Man using routinely collected macroinvertebrate water quality monitoring data and observed to expected (O/E) biotic index ratios derived from the BMWP system using RIVPACS [River InVertebrate Prediction And Classification System; Wright *et al.*, 2000]. Using site-specific physico-chemical data, RIVPACS predicts the macroinvertebrate assemblage and biotic scores expected if a river is unpolluted or undamaged and ratios of real and predicted biotic scores allow assessment of the biological quality of a site. The MacNeil *et al.* [2010] study, for the first time also examined the relationship between biocontamination and organic chemical water quality, specifically dissolved oxygen [DO], five-day biochemical oxygen demand [BOD_5)] and ammonia [Styles, 2001; MacNeil, 2006]. Applying the SBCI to the Isle of Man's rivers, 54% of sites exhibited no biocontamination, whereas 27% showed high or severe biocontamination. Considering all sites in the survey, there were significant negative relationships between the ACI and SBCI biocontamination metrics and O/E ASPT ratios [Figure 3], indicating that in sites where the invaders dominated, the biological water quality was lower, at least as reflected by biotic indices.

There was no significant relationship between O/E ASPT ratios and chemical (organic) water quality when either all river sites or only invaded river sites were considered, but in contrast, for uninvaded rivers O/E ratios responded significantly to changing chemical water quality, with biological water quality increasing as dissolved oxygen levels increased and ammonia levels decreased with no significant relationship evident for BOD_5. This could be due to the presence of pollution tolerant invaders undermining the responsiveness of the biotic indices to reflect changing water quality. The fact that some of the sites classed as having very good biological water quality by the BMWP/RIVPACS system also exhibited moderate to severe biocontamination, shows this aspect of assemblage structure is currently being missed in many monitoring programmes [MacNeil, 2006; MacNeil and Briffa, 2009]. This provides a strong argument that biocontamination should be regularly assessed alongside established biological and chemical monitoring programmes to provide an additional element 'missing' in current assessments of the 'true' ecological status of rivers.

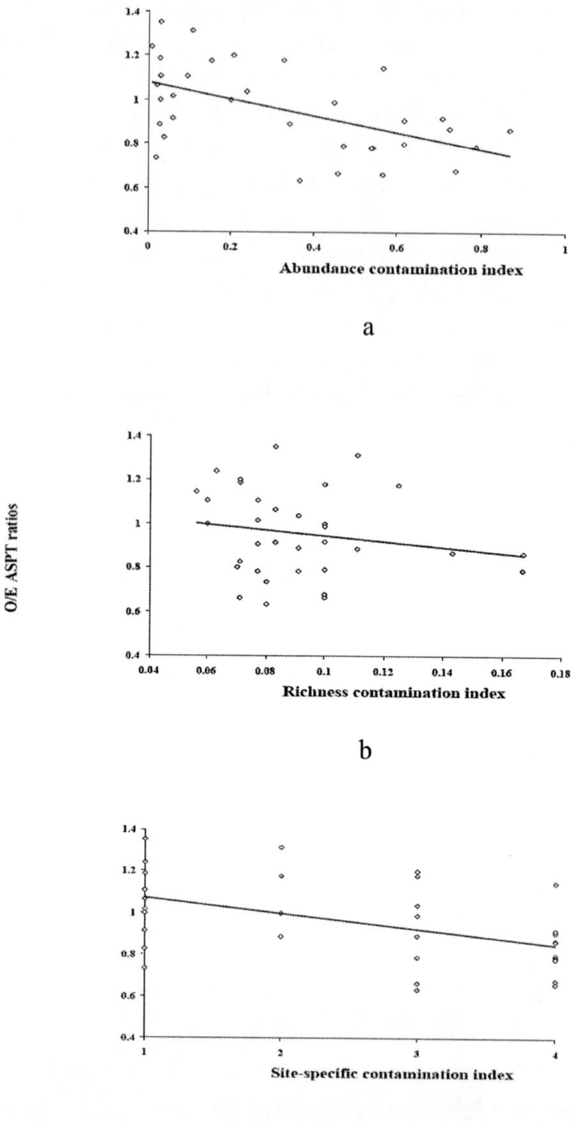

Figure 3. (a-c). Relationships between (a) abundance contamination index (ACI) (b) richness contamination index (RCIf) and (c) site-specific biocontamination index (SBCI) and observed/expected O/E ASPT ratios within invaded river sites (MacNeil and Briffa, 2009).

An obvious caveat to the usefulness of this biocontamination assessment method, is that when using family level macroinvertebrate data [as occurs with the BMWP system for the majority of time, when employed by the U.K. environmental protection agencies], some further taxonomic resolution will be required to identify invasive genera and species i.e. in order to designate contaminated families the operator will need the ability to distinguish between natives and invaders at the genus/species level. This is an important consideration [Gabriels *et al.*, 2005], as the extra time and level of expertise required will increase as the number of different invaders in the system increases.

Another point that should be borne in mind is that not all invaders have a similarly damaging impact on ecosystems and this has implications for assessing biocontamination, as biocontamination is not equivalent to 'biopollution' [Arbačiauskas *et al.* 2008]. It makes no assumptions regarding negative ecological or socio-economic impacts [Colautti and MacIsaac, 2004; Ricciardi and Cohen, 2007]. For instance, *C. pseudogracilis* unlike *Gammarus* spp., is mainly herbivorous, lives in marginal habitats and its introduction may even have supplemented the diets native fish populations [MacNeil *et al.*, 1999b].

CONCLUSION

Gabriels *et al.* [2005] recommended including invaders in indices such as the BBI, arguing that they were part of the community, despite the inherent dangers for native fauna. Similarly, Orendt *et al.* [2009] reviewed the role of invasive species in biological assessment and after considering the advantages and disadvantages of inclusion or exclusion of invaders in biological assessments of biodiversity and human impact, decided invasive species should be included. The inclusion of invasive species has the advantages of allowing analysis of the functioning of the whole ecosystem, as well as water quality functions and recording the taxonomy of the assemblage as a whole, can act as an early warning system that the assemblage may be about to change after the initial arrival of invaders [Orendt *et al.*, 2009]. For instance, where long-term biological data sets have been established alongside standard chemical measurements, it should be relatively simple to assess if the arrival of an invader has had a significant impact on the biotic indices, via the use of pre and post-invader data sets. However, Orendt *et al.* [2009] also noted that in the Netherlands, invaders are excluded from water quality assessments, indicating that different countries still handle the invader problem differently.

It is clear that the process of invasion cannot be halted [Strayer, 2010] and that invaders are expected to establish themselves as prominent parts of macroinvertebrate assemblages in an increasing number of European freshwaters [Orendt *et al.*, 2009; Strayer, 2010]. This latter point is reinforced by the very recent arrival of *D. villosus* in British waters [MacNeil *et al.*, in review]. Given the possible undermining of biotic indices outlined here, until there is a major overhaul of established biotic indices, it may be necessary to rely solely on chemical water quality and disregard biological water quality assessments in areas subject to invasion. At the very least, biological water quality monitoring within invaded freshwaters, should be treated with a far greater degree of caution than is currently the case.

REFERENCES

Andrew, T. E. (1998). Biological indications of the current status of rivers and lakes in Northern Ireland. In: J. G., Wilson (ED.), *Eutrophication in Irish waters*. Dublin: Royal Irish Academy, 48-54.

Arbačiauskas , K., Semenchenko, V., Grabowski, M., Leuven, R. S. E. W., Paunovic, M., Son, M. O., Csanyi, B., Gumuliauskaite, S., Konopacka, A., Nehring, S., Van der Velde, G., Vezhnovetz, V., and Panov, V. E. (2008). Assessment of biocontamination of benthic macroinvertebrate communities in European inland waterways. *Aquatic Invasions*, 3, 211-230.

Armitage, P. D., Moss, D., Wright, J. F., and Furse, M. T. (1983). The performance of a new biological quality score system based on macroinvertebrates over a wide range of unpolluted running water sites. *Water Research*, 17, 333-347.

Biological Monitoring Working Party (1978). *Final report: assessment and presentation of the biological quality of rivers in Great Britain*. December 1978. London: Water Data Unit, Department of the Environment.

Cardoso, A. C. and Free, G. (2008). Incorporating invasive alien species into ecological assessment in the context of the Water Framework Directive. *Aquatic Invasions*, 3, 361-366.

Clavero, M. and García-Berthou, E. (2005). Invasive species are a leading cause of animal extinctions. *Trends in Ecology and Evolution*, 20, 110.

Conlan, K. E. (1994). Amphipod crustaceans and environmental disturbance: a review. *Journal of Natural History*, 28, 519-554.

Colautti, R. I. and MacIsaac, H. J. (2004). A neutral terminology to define 'invasive' species. *Diversity and Distributions*, 10, 135-141.

Crawford, L., Yeomans, W. E. and Adams, C. E. (2006). The impact of introduced signal crayfish Pacifastacus leniusculus on stream invertebrate communities. *Aquatic Conservation: Marine and Freshwater Ecosystems*, 16, 611-621.

Den Hartog, C., Van den Brink, F. W .B. and Van der Velde, G. (1989). Brackish-water invaders in the River Rhine. A bioindication for increased salinity level over the years. *Naturwissenschaften*, 76, 80-81.

Dick, J. T. A. (2008). Role of behaviour in biological invasions and species distributions; lessons from interactions between the invasive Gammarus pulex and the native G. duebeni (Crustacea: Amphipoda). *Contributions to Zoology*, 77, 91-98.

Dick, J. T. A., Nelson, N. and Bishop, J. D. D. (1997). Introduction experiments with Gammarus spp. (Crustacea : Amphipoda) in the Isle of Man (British Isles), 1949 - 1995. *Journal of Zoology*, 242, 209-216.

Dick, J. T. A. and Platvoet, D. (1996). Intraguild predation and species exclusions in amphipods: the interaction of behaviour, physiology and environment. *Freshwater Biology*, 36, 375-383.

Dick, J. T. A., Platvoet, D. and Kelly, D. W. (2002). Predatory impact of the freshwater invader Dikerogammarus villosus (Crustacea: Amphipoda). *Canadian Journal of Fisheries and Aquatic Sciences*, 59, 1078-1084.

Ding, J., Mack, R. N., Lu, P., Ren, M., and Huang, H. (2008). China's booming economy is sparking and accelerating biological invasions. *Bioscience*, 58, 317-324.

European Commission (2000). *EC Water Framework Directive (2000/60/EC)*. Official Journal of the European Communities. Brussels: European Commission.

Gabriels, W., Goethals, P. L. M. and De Pauw, N. (2005). Implications of taxonomic modifications and alien species on biological water quality assessment as exemplified by the Belgian Biotic Index method. *Hydrobiologia*, 542, 137-150.

Gaston, K. J. and Spicer, J. I. (2001). The relationship between range size and niche breadth: a test using five species of Gammarus (Amphipoda). *Global Ecology and Biogeography*, 10, 179-188.

Hynes, H. B. N. (1950). Preliminary report on work on freshwater species of Gammarus in the Isle of Man. *Annual Report for 1949 (62) of the Marine Biological Station,* Port Erin, Isle of Man .

Hynes, H. B. N. (1954). The ecology of Gammarus duebeni Lilljeborg and its occurrence in freshwater in western Britain. *Journal of Animal Ecology*, 23, 38-84.

Hynes, H. B. N. (1955). Distribution of some freshwater Amphipoda in Britain. *Ver. Int. Ver. Theo. Ang. Limnol.* 12, 620-628.

Kelly, D. W., Bailey, R. J. E., MacNeil, C., Dick, J. T. A., and McDonald, R. A. (2006). Invasion by the amphipod Gammarus pulex alters community composition of native freshwater macroinvertebrates. *Diversity and Distributions*, 12, 525-534.

Kelly, D. W., Dick, J. T. A. and Montgomery, W. I. (2002). Predation on mayfly nymphs, Baetis rhodani, by native and introduced Gammarus: direct effects and the facilitation of predation by salmonids. *Freshwater Biology*, 47, 1257-1268.

Kelly, D. W., Dick, J. T. A., Montgomery, W. I., and MacNeil, C. (2003). Differences in composition of macroinvertebrate communities with invasive and native Gammarus spp. (Crustacea: Amphipoda). *Freshwater Biology*, 48, 306-315.

Kinzelbach, R. (1995). Neozoans in European waters – Exemplifying the worldwide process of invasion and species mixing. *Experimentia*, 51, 526-538.

Leuven, R. S. E., Van der Velde, G., Baijens, I., Snihders, J., Van der Zwart, C., Lenders, H. J. R., and Bij de Vaate, A. (2009). The river Rhine: a global highway for dispersal of aquatic invasive species. *Biological Invasions*, 11, 1989-2008.

MacNeil, C. (2006). *River Quality 2005*. Government Laboratory, Department of Local Government and Environment, Isle of Man Government. (available at www.gov.im/dlge/enviro/ govlabs).

MacNeil, C. and Briffa, M. (2009). Replacement of a native freshwater macroinvertebrate species by an invader: implications for biological water quality monitoring. *Hydrobiologia*, 635, 321-327.

MacNeil, C., Briffa, M., Leuven, R. S. E. W., Gell, F. R., and Selman, R. (2010). An appraisal of a biocontamination assessment method for freshwater macroinvertebrate assemblages; implications for current river water quality monitoring. *Hydrobiologia*, 638, 151-159.

MacNeil, C., Dick, J. T. A. and Elwood, R. W. (1997). The trophic ecology of freshwater Gammarus (Crustacea : Amphipoda); problems and perspectives concerning the functional feeding group concept. *Biological Reviews of the Cambridge Philosophical Society*, 72, 349-364.

MacNeil, C., Dick, J. T. A. and Elwood, R. W. (1999a). The dynamics of predation on Gammarus spp. (Crustacea : Amphipoda). *Biological Reviews of the Cambridge Philosophical Society*, 74, 375 - 395.

MacNeil, C., Dick, J. T. A. and Elwood, R. W. (2000). Differential physico - chemical tolerances of amphipod species revealed by field transplantations. *Oecologia*, 124, 1-7.

MacNeil, C., Dick, J. T. A., Gell, F. R., Selman, R., Lenartowicz, P., and Hynes, H. B. N. (2009). A long-term study (1949-2005) of experimental introductions to an Island; freshwater amphipods (Crustacea) in the Isle of Man (British Isles). *Diversity and Distributions*, 15, 232-241.

MacNeil, C., Elwood, R. W. and Dick, J. T. A. (1999b). Brown trout predation on native and introduced amphipods in N. Ireland, a behavioural study. *Ecography*, 22, 686 - 697.

MacNeil, C. and Platvoet, D. (2005). The predatory impact of the freshwater invader Dikerogammarus villosus on native Gammarus pulex (Crustacea: Amphipoda); influences of differential microdistribution and food resources. *Journal of Zoology*, 267, 1-8.

MacNeil, C., Platvoet, D., Dick, J. T. A., Fielding, N., Constable, A., Hall, N., Aldridge, D., Renals, T., and Diamond, M. (in review). *The Ponto-Caspian 'killer shrimp', Dikerogammarus villosus (Sowinsky, 1894), invades the British Isles.*

MacNeil, C., Prenter, J., Briffa, M., Fielding, N. J., Dick, J. T. A., Riddell, G. E., Hatcher, M. J., and Dunn, A. (2004). The replacement of a native freshwater amphipod by an invader: roles for environmental degradation and intraguild predation. *Canadian Journal of Fisheries and Aquatic Sciences*, 61, 1627-1635.

Orendt, C., Schmitt, C., Van Liefferinge, C., Wolfram, G., and De Deckere, E. (2009). Include or exclude? A review on the role and suitability of aquatic invertebrate neozoa as indicators in biological assessment with special respect to fresh and brackish European waters. *Biological Invasions,* 12, 265-283.

Panov, V. E., Alexandrov, B., Arbačiauskas, K., Binimelis, R., Copp, G. H., Grabowski, M., Lucy, F., Leuven, R. S., Nehring, S., Pauović, M., Semenchenko, V., and Son, M. O. (2009). Assessing the risks of aquatic species invasions via European inland Waterways: from concepts to environmental indicators. *Integrated Environmental Assessment and Management*, 5, 110-126.

Pinkster, S., Scheepmaker, M., Platvoet, D., and Broodbakker, N. (1992). Drastic changes in the amphipod fauna (Crustacea) of Dutch inland waters during the last 25 years. *Bijdragen tot de Dierkunde*, 61, 193-204.

Polis, G. A., Myers, C. A. and Holt, R. D. (1989). The ecology and evolution of intraguild predation: potential competitors that eat each other. *Annual Review of Ecology and Systematics*, 20, 297-330.

Ricciardi, A. (2001). Facilitative interactions among aquatic invaders: is an "invasional meltdown" occurring in the Great Lakes? *Canadian Journal of Fisheries and Aquatic Sciences*, 58, 2513-2525.

Ricciardi, A. and Cohen, J. (2007). The invasiveness of an introduced species does not predict its impact. *Biological Invasions*, 9, 309-315.

Ricciardi, A., Neves, R. J. and Rasmussen, J. B. (1998). Impending extinctions of North american freshwater mussels (Unionida) following the zebra mussel (Dreissena polymorpha) invasion. *Journal of Animal Ecology*, 67, 613-619.

Rosenberg, D. M. and Resh, V. H. (eds.) (1993). *Freshwater biomonitoring and benthic macroinvertebrates*. Chapman and Hall, New York.

Sala, O.E., Chapin, F. S., Armesto, J. J., Berlow, E., Bloomfield, J., Dirzo, R., Huber-Sanwald, E., Huenneke, L. F., Jackson, R. B., Kinzig, A., Leemans, R., Lodge, D. M., Mooney, H. A., Oesterheld, M., Poff, N. L., Sykes, M. T., Walker, B. H., Walker, M., and Wall, D. H. (2000). Global biodiversity scenarios for the year 2100. *Science*, 287, 1770-1774.

Solheim, A. L. and Gulati, R. D. (2008). Preface: 'Quantitaive ecological responses for the Water Framework Directive related to eutrophication and acidification of European lakes'. *Aquatic Ecology*, 42, 293-305.

Strange, C. D. and Glass, G. B. (1979). The distribution of freshwater gammarids in Northern Ireland. *Proceedings of the Royal Irish Academy*, 79, 145-153.

Strayer, D. L. (2010). Alien species in fresh waters: ecological effects, interactions with other stressors, and prospects for the future. *Freshwater Biology*, 55, 152-174.

Strayer, D. L. and Malcolm, H. M. (2007). Effects of zebra mussels (Dreissena ploymorpha) on native bivalves: the beginning of the end or the end of the beginning? *Journal of the North American Benthological Society*, 26, 111-122.

Styles, P. D. (2001). *River Quality 2000*. Government Laboratory, Department of Local Government and Environment, Isle of Man Government. (Available at www.gov.im/dlage/enviro/govlabs).

Tylianakis, J. M., Didham, R. K., Bascompte, J., and Wardle, D. A. (2008). Global change and species interactions in terrestrial ecosystems. *Ecology Letters*, 11, 1351-1363.

Van Riel, M. C., Van der Velde, G., Rajagopal, S., Marguiller, S., Dehairs, F. and Bij de Vaate, A. (2006). Trophic relationships in the Rhine food web during invasion and after establishment of the Ponto-Caspian invader Dikerogammarus villosus. *Hydrobiologia*, 565, 39-58.

Walley, W. J. and Hawkes, H. A. (1996). A computer-based reappraisal of the biological monitoring working party scores using data from the 1990 river quality survey of England and Wales. *Water Research*, 9, 2086-2094.

Wright, J. F., Sutcliffe, D. W. and Furse, M. T. (2000). *Assessing the biological quality of freshwaters – RIVPACS and other techniques*. Freshwater Biological Association, Ambleside.

In: Invasive Species
Editors: Joaquín J. Blanco and Adrian T. Fernandes
ISBN: 978-1-61942-761-7

Chapter 7

AQUARIUM TRADE AS A PATHWAY FOR THE INTRODUCTION OF INVASIVE SPECIES INTO MEXICO*

Roberto Mendoza Alfaro, Carlos Ramírez Martínez, Salvador Contreras Balderas, Patricia Koleff Osorio and Porfirio Álvarez Torres
Facultad de Ciencias Biologicas de la UANL, Monterrey, Nueva Leon, Mexico

ABSTRACT

The total ornamental fish industry worldwide (including dry goods) is valued at approximately US $15 billion dollars and it has been estimated that approximately one billion ornamental fish are commercialized every year with a value in the order of US $6 billion dollars. Freshwater species constitute the bulk of the trade; 90 percent of these are obtained from aquaculture and only 10 percent are wild captured. Around 800 to 1000 species and varieties are traded worldwide. Ornamental fish trade has been recognized as an important pathway for the introduction of non native species in several countries and present trends indicate that this pathway may turn into the main source of exotic invasive species in North America. Invasive species are characterized by posing different threats to the environment, the economy and human health. Among the main impacts provoked in the aquatic environment by invasive species are: competition with native species, hybridization, predation, introduction of diseases, habitat disruption and trophic webs modification. The introduction of exotic species has been related to the extinction of 54 percent of aquatic native species worldwide. Overall, 70 percent of the extinctions of North American fishes and 60 percent of those from Mexico are related to non native species, totally or partially. Aquarium trade has shown an accelerated increase during the last decade with a trade value of US $160 million dollars. This increase parallels the boost of exotic species in the country. In fact, in the 80's only 55 non native fish species were registered in Mexico and by 2004 the number raised to 118, of which 67 (58.26 percent) have turned invasive. Several facts contribute to explain this: i) The low amount of varieties cultured in Mexico (61 varieties pertaining to 19 species), compared to the

* A version of this chapter also appears in Freshwater Ecosystems and Aquaculture Research, edited by Felice De Carlo and Alessio Bassano, published by Nova Science Publishers, Inc. It was submitted for appropriate modifications in an effort to encourage wider dissemination of research.

huge number of varieties imported (more than 700 from 117 families), ii) The number of fish imported in Mexico; 40 million ornamental fish are traded annually, of which 45 percent are imported (nearly 18 million fish were imported in 2006) while 55 percent are captive bred iii) there is a lack of official regulations for the establishment and operation of farms producing ornamental fish and for the translocation of ornamental fish within the country. As a result, ornamental fish species have been established in 9 out of 10 continental aquatic regions of Mexico. Some of these species have already severely impacted the environment and the economy in most regions of the country.

Introduction

The Ornamental Fish Industry Worldwide

The global ornamental fish industry (including dry goods) is valued at approximately US $15 billion dollars (Bartley, 2000) and currently about one billion ornamental fish are traded anually (Whittington and Chong, 2007), with a value of US $6 billion dollars (Holthus and Gamain, 2007) of which Asia provides 78 percent of such exportations worldwide (UNEP, 2008).

The main trading countries obtain most of their product from developing countries such as Philippines, Colombia, Peru, Brazil and Sri Lanka to later be re-exported to the rest of the world. However, the ornamental fish trade is not just important in terms of its share in international trade. The sector is frequently a welcome provider of employment opportunities (Wijkström, 2002) and an important source of income for these economically depressed rural, coastal and insular communities where this industry manage to account under the proper developmental scenario, up to 60 percent of the local economy as in the case of the Amazon (Chao and Prang, 1997; Dowd and Tlusty, 2000; Huanqui-Canto, 2002).

Most of the production of cultured ornamental fish focuses on freshwater species. Approximately 90 percent of freshwater ornamental fish are captively bred, compared to only 10 percent collected in the wild (Dawes, 1998, Chao and Prang, 2002; Wabnitz et al., 2003). This contrasts with marine ornamental species which are mostly (99 percent) collected in the wild and just 1 percent are reared (Hershberger, W.K., 2003).

The variety of fish involved in this industry is very significant. The estimated number of species artificially reared in the freshwater ornamental trade range from 800 to more than 1,000 species, despite the scarce studies existing concerning their culture (Hernández, 2002; Tlusty, 2002). This is an important figure when it is compared to the total number of finfish species used for food or stock enhancement in commercial aquaculture which is estimated to be only about 180 (Williams 1997). In contrast, it is estimated that a total of about 2,600 species are harvested in industrial and artisanal fisheries. (Hershberger, 2003).

The Ornamental Fish Industry as a Pathway for Aquatic Invasive Species

There is a clear and direct relationship between international trade and the spread of invasive alien species (Burgiel., et al., 2006) and the most important pathway for non-indigenous introductions into North America has been intentional or unintentional importations of organisms associated with international trade (Rixon et al., 2005). In this regard, at the international level, aquarium trade has been widely recognized as an important pathway for the introduction of alien aquatic species into new environments (Taylor et al., 1984; Welcomme, 1992) and the current trends indicate that it may become soon the main source of introductions of invasive exotic species in North America (Courtenay, 1995; Courtenay and Williams, 1992). Following the criteria Burgiel and co-workers (Burgiel., et al. 2006), several facts support this contention:

- The rate of introductions of invasive alien species likely correlates with the volume of trade. More than one billion ornamental fish are traded every year (Whittington and Chong, 2007)
- More introductions lead to a greater probability that an invasive alien species will become established. The rate of movement of species between countries has accelerated since 1900 (Welcomme, 1992). From approximately 100 species of ornamental fish that have been recorded as introduced into USA and Canada natural waters via the aquarium trade, 40 have established populations (Rixon et al., 2004), while in the case of Mexico from 115 introduced exotic species 67 have become established (Contreras, In Press).
- An increasing variety of goods and means of transport increases both the potential array of species that may be moved and their pathways for transfer. Only the state of Florida in the US produces 800 varieties of freshwater fish (Tlusty, 2002) and it has been stated the high potential risk posed by the tremendous taxonomic diversity of the ornamental fish industry (Weigle et al., 2005). On the other hand, ornamental fish are transported by planes, ships and trucks all over the world.
- More frequent delivery of goods from and to a wider range of countries and habitats increases the rate and variety of potential introductions. Data from the last decade show that 133 countries imported ornamental fish, while 146 countries exported them (Huanqui-Canto, G, 2002).
- Faster modes of transport may improve an organism's chance of survival while in transit. Fuller (In press) points out that aquarium trade has become the most important pathway in the introduction of invasive species in the Gulf of Mexico and South Atlantic region and that potential invasive species are more likely to survive their journey now than in the past because the increased speed of travel and better packaging techniques.

Also, internationally, ornamental fish farms constitute one of the main pathways for the introduction of exotic species into the natural environment. This is in part due to the high number of continuous changing species and varieties of fish produced and the closeness to natural freshwater bodies (Copp et al., 2005). Fish are known to have escaped or were purposefully released from aquarium fish culture facilities, and others were introduced by well-meaning but misguided aquarists. The escape or release of aquarium fishes into open waters has never resulted in beneficial introductions. In many such instances, there have been negative impacts on native fishes and habitats (Contreras-Balderas and Escalante, 1984; Courtenay and Stauffer, 1990).

The Ornamental Fish Industry in Mexico: History, Size and Scope

The ornamental fish industry in Mexico initiated in the 50's, when ornamental fish were reared for the first time and important exhibitions in public aquaria were mounted. It is also at this time, that the first ornamental fish production and commercialization association was established (Ortiz, 1997).

However, the development of the industry through the time was incipient, as stated by the fact that during the 70's there were only 5 producers and 100 retailers. This industry started culturing cyprinids, brought from Asia by the federal government, as an alternative to agriculture. The value of the industry at the time was US $500,000 dollars, while the importations, represented only US $21,000 dollars (INP, 1974). It is worthy to point out that during the 70's there were severe restrictions to import exotic species from Asia and Africa and the availability of equipment for tanks was minimal.

The effort to culture ornamental fish continued and during the 80's several fish farms, originally created to culture tilapias and freshwater prawns, were reconverted to culture ornamental fish due to former cultured species low yields. However, the growth of the industry kept on being low not only because the government stopped subsidizing the production, but also due to the lack of experience and technical skills, together with the poor existing sanitary control measures at that time. (SEPESCA, 1984).

During the early 90's the demand increased and the federal government began providing proper financial support to establish farms. However, the most important contribution to the industry development and expansion was the nation economic crisis during the mid 90's, which limited the importers and pushed local producers to sustain the market with fish accessible to consumers.

In the same sense, from the official records of imported fish and those of registered farmers in the states of Morelos, Yucatán, Hidalgo and Puebla, it can be assessed that ornamental fish commerce grew more than 100 percent in the past decade (Figure 1) responding to the continuous growing demand, particularly in the three main large cities (México City, Guadalajara and Monterrey), where owing to urbanization and lack of space fish rapidly became the preferred pets.

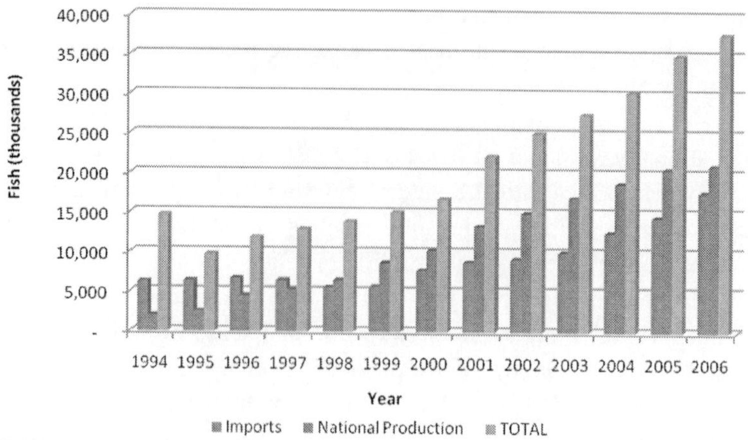

Source: INEGI 1994-2006 and Annual Fisheries Statistics 1994-2004.

Figure 1. Sales of ornamental fish in México in the period of 1994-2006.

According to information on the Annual Statistical Reports from Foreign Commerce of Mexico (1994-2007), and the Annual Fisheries Statistics (1994-2003) the fresh water aquarium fish trade in México, increased more than 100 percent, going from approximately 13 million fish to nearly 37.5 million from 1994-2006, representing thus an annual growth rate of 8 percent (Figure 1).

However, accordingly to a spoken statement from five of the main aquarium fish traders from México City, it is a common practice to declare less than the amount of fish specimens than are actually received in shipments, leading to the inaccuracy of official reports. In this way, importations may be underestimated by almost 25 percent. This situation may indicate that the number of freshwater fish species traded in México may actually be a bit above 40 million individuals a year (Figure 2).

Approximately 45 percent of these fish are imported, while the remaining 55 percent are raised in ornamental fish farms located around several states of the country. These farms are the source of employment to more than 1,250 persons (Ramírez and Mendoza, 2005). According to the last economical census there are 5,126 aquarium stores (INEGI, 2005), however unofficial figures point to the existence of 20,000 aquarium stores, employing around 41,000 persons.

Ornamental freshwater fishes generate sales accounting to US $161,500,000 dollars at retail prices, of which 51 percent comes from importations and 49 percent were provided by national production, with an average annual increase of 18 percent in the last 12 years (Figure 3)

As a consequence of freshwater ornamental fish industry rapid growth, diverse risks have also increased. Unfortunately, a large amount of those exotic organisms accidentally and/or intentionally have been released to natural aquatic environments and many of them became invasive.

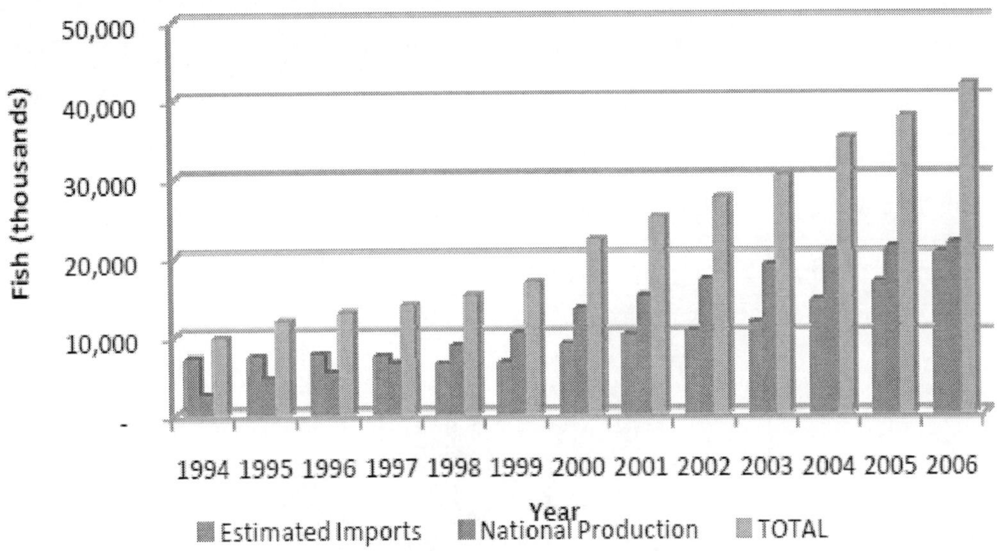

Figure 2. Estimated sales of fresh water aquarium fishes in México, during 1994-2006.

National Production

The bulk of the national production is concentrated in the state of Morelos, which would be the equivalent of the State of Florida in the USA.

The number of ornamental fish farms at the national level is uncertain, but it is estimated to be around 250. To find official figures a request to the National Institute of Access to Information (IFAI) was done, and the National Commission of Aquaculture and Fisheries answered that "...according to the Law of Transparency there is not a systematized register of ornamental fish production or producers, however there would be a gross estimate of 128 farms in the state of Morelos..." Nevertheless according to the SAGARPA Coordinator of Fisheries Promotion at Morelos State there would be at least 200 farms. Moreover, from a survey made at the national level there are farms located in other 20 states (Figure4), but these would not surpass 50. Most of these farms are rustic and are operated at the family level.

The state of Morelos produces around 17 million individuals/year (CONAPESCA, 2005). These farms are concentrated in 17 of the 33 municipalities of the State and their size varies from 2,000 m^2 to 1.5 has. Most of them (85 percent) are considered small (between 2,000 to 40,000 m^2).

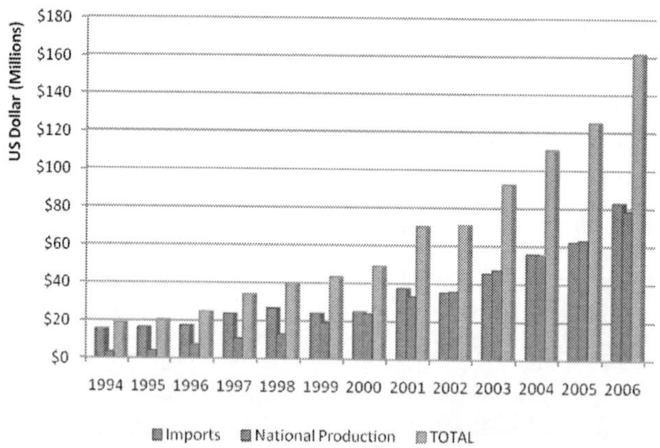

Figure 3. Retail amount of sales from ornamental freshwater fish in México during the period 1994-2006.

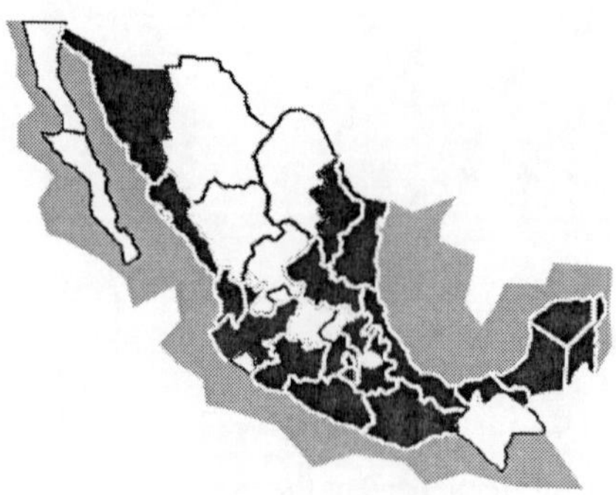

Figure 4. Mexican states where facilities for the culture of ornamental freshwater fish are located (marked in black).

The majority of the ponds are concrete made, but the larger ones are rustic earthen ponds (Figure 5). The bulk of these fish production is sold mainly in Mexico City, Jalisco, Nuevo León, Baja California and Guanajuato States. Fish are sold directly in the farms or sent to the wholesalers market in Mexico City (SAGARPA, 2005).

Nearly all the producers are organized in two main associations. Of the 128 farms officially registered, 45 belong to the APPOEM Ornamental Fish Farmers Association of Morelos (Asociación de Productores de Peces de Ornato del Estado de Morelos) and 54 to "La Perla" (Productores de Peces de Ornato "La Perla" de Morelos A.C) however there are 29 independent farms. The first of these associations was created under an initiative of ornamental farmers, while the second was created as a request from the state government, so the producers could benefit from the different government support programs and economic subsidies. Ornamental fish farmers have benefited from government official programs not only by obtaining significant tax discounts, but also by being eligible to financial support and subsidies of up to US $7,000 dollars, only if their farms are officially registered. Nowadays, besides representing the producers they serve as a discussion forum to deal with different problems such as the adaptation of their facilities to new regulations, and homogeneity of prices to face intermediaries. The greatest part of the production takes place from March to November, as the winter temperatures are rather low for the production of goldfish which have the highest demand. One of the producers' associations (La Perla) has been looking for technical assistance of the Morelos State University to increase their production levels. The main problem faced by producers is the supply of good quality broodstock and the control of genetic quality of the different varieties of fishes produced.

The recent success of the ornamental fish industry has attracted the interest of several entrepeneurs from other sectors, willing to invest on this activity, which will be reflected in its growth, in the short term.

The variety of ornamental fish produced in Mexico can be considered low when compared to the international standard, as there are only 61 varieties from 19 species currently reared. In contrast, it has been estimated that nearly 700 varieties pertaining to 117 families are imported each year into the country (Alvarez-Jasso, 2004).

Figure 5. Typical ornamental fish farm in the state of Morelos, Mexico.

However, the economic accessibility to these varieties is restricted and contributes to explain the drop in consumption during the times of economical crisis.

The varieties currently reared are: Livebearers (mollies, guppies, platies and swordtails) constituting the bulk of the production, however many usual oviparous fish such as cichlids (angels and different African cichlids), anabantids (gouramies and betas), characins (tetras), silurids (catfish, corys and plecos) and chiefly a vast array of goldfish and other cyprinids (barbs, rasboras, sharks) are also cultured.

Importations

The ornamental fish industry in Mexico has been operating for more than 50 years and although at the beginning, this industry depended mostly on species captured from the natural environment and some imported from Central and South America, imports, for year 2006 have been estimated in nearly 18 million fish.

The main reasons to import fish into Mexico are:

- The high retailers' price of several alien species
- The quality of many imported species, that can't be attained by national producers
- The variety of organisms, which is a fundamental aspect, since the demand in the ornamental fish industry is fashion-based
- The availability of many species during the whole year, while in Mexico some species are just available during a part of the year (e.g. goldfish)
- The support of the academic sector from exporting countries

In the case of importations coming from South America, especially from the Amazons region pertaining to Brazil, Colombia and Peru, a significant change in the volume of fish exported has been noticed during the last five years. Not only has the total number of individuals increased in an important way, but also the participation of Colombia over those of Brazil and Peru has increased over time. This is in great part due to the Treaty on Free Trade Between the Republic of Colombia, the Republic of Venezuela and Mexico also known as G3. In particular, this treaty not only eases the importations of goods coming from Colombia, but also gives this country an important advantage over the rest due to the tariff elimination (Figure 6).

On the other hand, Almenara (2001) points out a disadvantage between importers in relation to the NAFTA (North American Free Trade Agreement), signed between Mexico, Canada and the United States of America. According to this agreement, aquatic ornamental organisms whose origin is either Canada or the USA are duty free, in comparison with 23 percent import duty from other countries; this also applies to the duty on freight and packing paid to transport them. As all fishes coming from Asia pass through the USA, this allows some cheating importers to turn China goldfish into 'American bred' goldfish and, thus end up paying less than US $15.00 dollars in customs fee per shipment, compared to the more than US $2,500.00 dollars per shipment with legal papers that honest importers pay. The lack of professional staff at the Customs Department cannot control this tax and duty evasion that is harmful to the country. Since it is virtually impossible to determine the precise origin of tank-bred species of fish, the Customs authorities rely only on the exporter's paperwork.

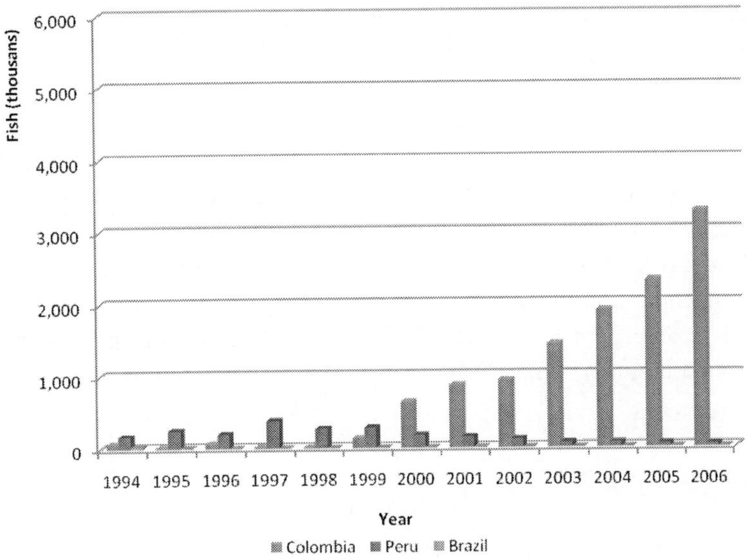

Figure 6. Importations of ornamental freshwater fish from South America to México during 1994-2006.

Within this context, in spite of the substantial increase in commercial exchange between Mexico and the USA after the NAFTA treaty establishment, more than ten years ago, it is interesting to note that ornamental fish importations follow a cyclic pattern in which every six years when the federal government changes (1994, 2000, 2006) there is an economic constriction, which coincides with the drop of importations. Notwithstanding, importations remain to be the second source of ornamental fish in Mexico.

The more frequent imported freshwater fish families are:

Table 1. Imported families of ornamental fish according to the number of species and number of individuals

Families	Species	Families	Number of individuals
Cichlidae	107	Characidae	1.187.788
Characiidae	64	Cyprinidae	1.092.506
Cyprinidae	27	Osporonemidae	689.776
Callichthyidae	24	Poeciliidae	456.563
Loricariidae	20	Cichlidae	161.276
Conaitidae	15	Ambassidae	133.003
Aplocheilidae	13	Callicthyidae	130.385
Anabantidae	12	Pangasiidae	118.739
Pimelodidae	11	Loricariidae	102.743

The importance of these families changes when the number of individuals per family is taken into account. This shows that in some cases, such as the cichlids, a relatively high number of species, but a small number of individuals is imported, which contrasts with characids in which a high number of individuals and a low number of species is imported, obeying to the market demand.

It has been suggested that the risk of establishment largely depends on the biogeographic origin of the organisms (Ramírez y Mendoza, 2005). Thus, many fish imported from Asia, Europe and the USA are actually captively bred and consequently are less prone to survive in the wild; however the threat associated to the introduction of these organisms are diseases caused by exotic pathogens and parasites. In contrast, most of those organisms imported from South America, particularly from the Amazons, are captured in the wild and consequently represent a high risk of establishment in a similar environment, as has been the case with some cichlids and fish of the Loricariidae family. In relation to this, it should be noticed that ornamental fish importations from South America have been increasing, while those from Florida are decreasing, due to import substitutions made by domestic breeders (Almenara, 2001). However, a large amount of fish has been historically imported from the USA.

Problems of the Industry

During the last decade the culture and commercialization sectors of the ornamental fish industry have been experiencing some difficulties due to the following aspects:

- Lack of promotion
- No research support
- Lack of internal organization
- Lack of an efficient regulation

The industry lacks a real and reliable organization due to fierce competition between and among producers, importers, wholesalers and retailers ("changarreros"). To this should be added the scarce research on different aspects of ornamental fish species (particularly on reproduction, nutrition, pathology and genetics), the lack of communication (only a minority gathers in an annual congress), lack of training or information within the context of sustainability and responsibility towards the environment (there are very few national aquarium magazines). The official figures and statistics concerning the sector (number of producers, areas of production, volume and origin of imports or countries where fish are exported) are inaccurate and often non trustworthy.

As a consequence of the above mentioned problems, a continuous negative impact to the environment has been registered, particularly due to the absence of a specific and efficient legal framework. Indeed, the absence of regulatory measures requesting farmer owners to have an adequate infrastructure has originated the continuous escapes of native or exotic species into the wild. An example is the state of Morelos, where most of ornamental fish farms are concentrated. In the rivers of this state there are a total of 22 species registered, 8 of which are native while 14 are exotic (Contreras-MacBeath, 1998).

In addition, several voids in the environmental laws together with the lack of necessary enforcement have given place to illegal trade (importation and exportation) and translocation within the country of these species. In fact, despite the existence of a seven day quarantine period prior to selling the product after importation, in many cases the law is not observed by non-ethical people whose businesses are based on just moving their fishe rapidly without caring for quality. The results of these improper practices are disastrous to those importers who have been investing their money, effort and time in order to offer healthy organisms according to the prevailing rules. They are at a total disadvantage against those people who keep on cheating the law and doing whatever possible to keep giving away their products without any consideration. An undesirable, yet expected consequence is that all importers are pushed to ignore the quarantine period as well. Even though the authorities know that the law is not being properly observed (or not observed at all), they cannot act, due to the lack of resources to enforce the law. In consequence, this leaves the legal importers sector abandoned at the mercy of unscrupulous merchants (Almenara, 2001). To this should be added the difficulty to obtain the permits, the frequency to get them (every three months), the strict specificity of permits (which increase the price with each new species) and the geographical distance from Mexico City to Mazatlan city (1,085 Km) to obtain the permit where fisheries authorities are based.

Ecological Risks

In Mexico, fish are commonly released by farmers when prices drop or when a disease is detected, because due to the cost of antibiotics they will not obtain the desired commercial value. This situation implies a high risk of escapes generating important problems, such as competition for food resources and space, predation, native species displacement and alteration of communities' trophic structure, hybridization, incidence of diseases and parasites and habitat alteration, (Contreras-Balderas, 1969, Contreras-Balderas and Escalante, 1984, Lassuy, 2000; MIT Sea Grant, 2002; Goldburg, 2001; Hopkins, 2001).

At the national level, most of the ornamental fish farms lack of adequate biosecurity systems, implying the risk of disease propagation and escapes of native and non-native species. In general, operating farms use extensive culture systems, with low or non-existent technology and are operated by untrained personnel (Ramírez, 1999).

Furthermore, most of the farms are not constructed under the official standards and regulations established by the environmental authority, such as conducting environmental impact assessments to demonstrating that the farm will not pose any threats to the environment, as stated by the environmental law (DOF, 2000) and often farms do not adopt current sanitary regulations, (NOM-010-PESC-1993, NOM-011-PESC-1994). As a result, many farms are located in places where they represent a high ecological risk and are operating with production systems far from being adequate.

This situation is critical as Mexico is one of the five most diverse countries in the world and among the 12 considered as megadiverse (CONABIO, 2006), that hosts 60 and 70 percent of the known biological diversity in the planet. This high biodiversity is due in part to the biogeographic conditions of the country, in a transitional zone between the two large neotropical and neartic regions. As a result, Mexico has a rich source of flora and fauna and particularly a high species diversity of freshwater fish. There are more than 47 families, composed by approximately 506 species, 375 of which are exclusively restricted to freshwater environments (Contreras-Balderas, et al., 2008; Torres-Orozco, y Kobelkowsky, 1991). Additionally to this high diversity, the important variety and isolation of the different basins has given place to several endemic species which ecological importance is even higher.

One of the reasons why the impact of alien invasive species has passed inadvertently in Mexico is the high biological diversity of the country. In this regard, it has been stated that because some ecosystems are richer in number of species than are others, it is likely that such ecosystems would be less hospitable to establishment and invasion by nonnative species through species packing over time (Courtenay, 1990). In the same sense Moyle and Light (1996) mentioned that major community effects of invasions are most often observed where the number of species is low and that invaders are most likely to extirpate native species in aquatic systems with extremely low variability.

Unfortunately, it should be pointed out that according to the National Fishery Chart (DOF 2004) 9 of the 10 aquatic provinces of the country host several species that have been introduced due to aquarium trade.

Some of these exotic species have reached critical areas such as Mexico's Natural Protected Areas, characterized by a high endemism, e.g. African jewel cichlid in Cuatro Cienegas Basin (Contreras-Balderas and Ludlow, 2003). Unfortunately, according to Contreras-Balderas, (1971) and Hamilton (2001) native species, especially indigenous species that evolved in isolated ecosystems are often very poorly equipped to survive once invasive alien species are introduced. Additionally, most of the endemic fishes are small and therefore, subject to more adverse impacts through introductions of small bait or ornamental fishes than with larger fishes (Courtenay and Deacon, 1982). The prediction of an invasion by non-native fishes in Cuatro Cienegas and its consequences was reported since 1983 (Courtenay, 1983).

There are two reasons to expect invasion threats from the aquarium industry to increase with time. Firstly, the pool of potential invaders is ever expanding as the industry searches for new, potentially popular species to market. Secondly, because most aquarium species are of tropical and subtropical origin, the probability of their establishment in North America will increase with climatic warming (Rixon et al., 2004).

Finally, it should be noted the tight relationship existing between a highly deteriorated aquatic ecosystems, as those of the state of Morelos, and the establishment of invasive species as the plecos or the convict cichlid. With this regard, it is difficult to estimate to which extent the severe pollution was the direct cause of extinction of native species and what role was played by the invasive species. Anyhow, it has been reported that most impacts, to date, have been in areas where habitats modification has already stressed fish populations (Minckley and Deacon 1968; Deacon 1979; Courtenay, et al, 1985; Welcomme, 1988; Courtenay and Robins, 1989). Moreover, the environmental tolerance of non-indigenous fish combined with increasing habitat disruption in streams and lakes assures their continued dispersal into formerly unoccupied areas (Boydstun, et al, 1995). This is also true regarding their establishment, as in aquatic systems with intermediate levels of human disturbance, any species with the right physiological and morphological characteristics can become established and long-term success (integration) of an invading species is much more likely in an aquatic system permanently altered by human activity than in a lightly disturbed system (Moyle and Light, 1996).

Conclusion

Several aspects in the aquarium trade industry in Mexico need urgently to be reviewed and modified from current trade and farming practices to those related to norms, standards and regulations applicable to the industry. Among these the need to adequately regulate importations and adopt preventive measures such as risk analysis and the elaboration of white and black lists. In the case of production there is a strong need of biosecurity standards and preventive measures such as HACCP (Hazard Analysis to Control Critical Points) plans for each farm. While in the case of commercialization international measures such as those proposed by the Ornamental Aquatic Trade Association should be adopted.

REFERENCES

Almenara Roldán, S. 2001.The Current Status of the Aquarium Industry in Mexico. *OFI Journal Issue* 35. 3pp

Alvarez-Jasso, M. 2004. "La introducción de peces ornamentales en México a través de las importaciones durante el año 2001 y su ordenamiento. *Tesis. De Licenciatura*. Facultad de Ciencias. UNAM. México. 75 p.

Bartley, D. M. 2000. Responsible ornamental fisheries. *FAO Aquat. Newsl.* 24, 10–14.

Boydstun, Ch., Fuller, P. and Williams, J. D. 1995. Nonindigenous Fish. In: Our living resources. *A report to the nation on distribution, abundance, and health of U.S. Plants, Animals, and Ecosystems.* U.S. Department of the Interior. National Biological Service. USGS, Florida. USA.

Burgiel, S., G. Foote, M. Orellana and A. Perrault. 2006. Invasive Alien Species and Trade: Integrating Prevention Measures and International Trade Rules. *Center for International Environmental Law and Defenders of Wildlife.* 54pp

Chao, N. L., Prang, G., 1997. Project Piaba—towards a sustainable ornamental fishery in the Amazon. *Aquarium Sci. Conserv.* 1, 105–111.

Chao, N. L., Prang, G., 2002. Decade of project Piaba: reflections and prospects. *OFI Journal* (39):1-9.

Conabio 2006. Capital natural y bienestar social. *Comisión Nacional para el Conocimiento y Uso de la Biodiversidad*, México. 70pp

Conapesca (Comisión Nacional de Acuacultura y Pesca). 2005. Respuesta a la consulta número 00008000050305 del Instituto Federal de Acceso a la Información Pública. Documento Oficial.

Contreras-Balderas, S. and M. A. Escalante. 1984. *Distribution and Known Impacts of Exotic Fishes in México*. Chapter 6.102=130. In: Distribution, and Management of Exotic Fishes. Eds. W. R. Courtenay y J. R. Stauffer. John Hopkins University Press

Contreras-Balderas, S. 1971. El hombre, el desierto y la conservación de los recursos naturales. *Revista Fauna Silvestre y Cacería*, 3:20-24./ Armas y Letras,(Universidad de Nuevo León), 1974(1):43-50.

Contreras-Balderas, S. 1969. Perspectivas de la ictiofauna en las zonas áridas del Norte de México. Mem. Primer Simp. *Internacional de Aumento de Producción de Alimentos en Zonas Áridas*. ICASALS, Texas Tech. Publ. 3:294-304

Contreras-Balderas, S. En Edición. Peces Exóticos Invasivos de agua dulce en México. En trámite de impresión.

Contreras-Balderas, S. y Ludlow, A. 2003. Hemichromis guttatus, Nueva Introducción en Cuatro Ciénegas, Coahuila, Primer Reporte para México (Pisces: Cichlidae). *Vertebrata Mexicana*, 12:1-5.

Contreras-Balderas, S. y Medina Gándara, J. 1971. Ecología de Zonas Áridas de México, Para: Inventario de Recursos Humanos, Materiales y Financieros de Ciencia y Tecnología para Zonas Áridas. Informe Consejo Nacional de Ciencia y Tecnología, México.

Contreras-Balderas, S., G. Ruiz-Campos, J. J. Schmitter-Soto, E. Diaz-Pardo, T. Contreras-McBeath, M. Medina-Soto, L. Zambrano-Gonzalez, A. Varela-Romero, R. Mendoza-Alfaro, C. Ramirez-Martinez, M. A. Leija-Tristan, P. Almada-Villela, D. A. Hendrickson, and J. Lyons, 2008. Freshwater fishes and water status in Mexico: a country-wide appraisal. *Aquatic Ecosist. Health and Managmnt.*, 11(3):246-256.

Contreras-MacBeath, T. H. Mejia Mojica and R. Carrillo Wilson. 1998. Negative impact on the aquatic ecosystems of the state of Morelos, Mexico from introduced aquarium and other commercial Fish. *Aquarium Sciences and Conservation,* 2: 67-78.

Copp, G. H., K. Wesley and L. Vilizzi. 2005. Pathways of ornamental and aquarium fish introductions into urban ponds of Epping Forest (London, England): the human vector. *J. Appl. Ichtyol.* 21:263-274

Courtenay, W. R. Jr. 1983. Fish introductions in the American Southwest: a case history of Roger Springs, Nevada. *The Southwestern Naturalist* 28:221-224.

Courtenay, W. R., Jr,. 1990. Fish conservation and the enigma of introduced species. *Proc. Aust. Soc. Fish Biol.* 8: 11-20

Courtenay, W. R., Jr,. 1995. The case for caution with fish introductions. *American Fisheries Society Symposium* 15: 413-424.

Courtenay, W. R., Jr, J. E. Deacon, D. W. Sada, R. C. Allan, and G. L. Vinyard. 1985. Comparative status of fishes in along the course of the pluvial White River, Nevada. *Southwestern Naturalist* 28:221-224.

Courtenay, W. R., Jr. and J. E. Deacon .1982. Status of introduced fishes in certain spring systems in Southern Nevada. *Great Basin Naturalist* 42:361-366.

Courtenay, W. R., Jr, and C. R. Robins. 1989. Fish introductions: good management? *Critical Reviews in Aquatic Science* 1: 159-172.

Courtenay, W. R., Jr, and J. R. Stauffer, Jr. 1990. The introduced fish problem and the aquarium fish industry. *Journal of the World Aquaculture Society* 21: 149-159.

Courtenay, W. R., Jr. and J. D. Williams. 1992. Dispersal of exotic species from aquaculture sources, with emphasis on freshwaters fishes. In: A. Rosenfield and R. Mann, eds. Dispersal of living organisms into aquatic ecosystems. *Maryland Sea Grant College, College Park.* pp 49-81

Dawes, J., 1998. International experience in ornamental marine species management. Part 1: perspectives. *Ornamental Fish Int. J.,* 26, February 1999. http://www.ornamental-fish-int.org/marinespecies1.htm.

Deacon, J. E. 1979. Endangered and threatened fishes of the West. In: The endangered species: a symposium. *Great Basin Naturalist Memoirs* 3: 41-64.

DOF (Diario Oficial de la Federación). 2000. Ley General del Equilibrio Ecológico y Protección al Ambiente. *Ultima Reforma. 7 de Enero*, 2000. México.

DOF (Diario Oficial de la Federación). 2004. Carta Nacional Pesquera. México

Dowd, S., Tlusty, M. F. 2000. Project Piaba—working toward a sustainable natural resource in Amazon freshwater fisheries. *Endangered Species Update 17, Univ. MI. School Natural Resources*. 88– 90.

Fuller, P. In Press. Nonindigenous aquatic pathways analysis for the Gulf and South Atlantic region. *Report produced for the Aquatic Nuisance Species Taskforce*. 55 pp (In press)

Goldburg, R. J., M. S. Elliott, R. L. Taylor. 2001. Marine Aquaculture in the United Status: *Enviromental IMpacts and Policy Options*. Pew Oceans Commission, Arlington, Virginia

Hamilton, L. A. 2001. Invasive alien species: a major continuing threat to the health and integrity of freshwater ecosystems. *Resource Futures International,* Ottawa. (9): 1-9.

Hernández, O. D. 2002. Producción de Peces de Ornato. *Hypatia*. (4). http://hypatia.morelos.gob.mx/no4/peces ornato.htm

Hershberger, W. K. 2003. Aquaculture: Its Role in the Spread of Aquatic Nuisance Species. *National Center for Cool and Cold Water Aquaculture*. USA. 13 pp

Hopkins, C. C. E. 2001. *A review of introductions and transfers of alien marine species in the North Sea area*.

Holthus, P. y Gamain, N. 2007. Del arrecife al minorista. *Conservación Mundial.* http://www.iucn.org/publications/worldconservation/docs/2007_01/12_world_conservation_2007_01_es.pdf

Huanqui Canto, G., 2002. *El comercio mundial de peces ornamentales*. Prompex. Perú. 10 p.

INEGI (Instituto Nacional de Estadística, Geografía e Informática). 2001.. Censos Económicos 1999. Resultados Definitivos. México.

INEGI (Instituto Nacional de Estadística, Geografía e Informática). 2005_a.. Censos Económicos 2004. Resultados Generales. México.

INEGI (Instituto Nacional de Estadística, Geografía e Informática). 2005_b.. Anuario Estadístico del Comercio Exterior de los Estados Unidos Mexicanos. 2004.

INEGI (Instituto Nacional de Estadística, Geografía e Informática). 2005_c. Anuario Estadístico del Comercio Exterior de los Estados Unidos Mexicanos 2004.

INP 1978. Estado de la acuicultura en México. En: la acuicultura en América Latina. Instituto Nacional de la Pesca Mexico y FAO, Informes de Pesca, 159, Vol. 3: S/C 13

Lassuy, D. R., 2002, Introduced Species as a factor in extinction and endangerment of native fish species. *Workshop: Management, implications and co-occurring native and introduced fishes proceedings,* Portland Oregon: 27-28,

Minckley, W. L. and J. E. Deacon. 1968. Southwestern fishes and the enigma of "endangered species" *Science* 159: 1424-1432.

MIT Sea Grant,. 2002. Exotic species an ecological roulette with nature. MIT Sea Grant College Program, *Costal Resources. Fact sheet.*

Moyle, P. B. and T. Light. 1996. Biological invasions of fresh water: empirical rules and assembly theory. *Biological Conservation* 78: 149-161.

NOM-010-PESC-1993. Norma Oficial Mexicana que establece los requisitos sanitarios para la importación de organismos acuáticos vivos en cualesquiera de sus fases de desarrollo, destinados a la acuacultura u ornato, en el territorio nacional. *Diario Oficial de la Federación*. 20 de Julio, 1994.

NOM-011-PESC-1993. Norma Oficial Mexicana para regular la aplicación de cuarentenas, a efecto de prevenir la introducción y dispersión de enfermedades certificables y notificables, en la importación de organismos acuáticos vivos en cualesquiera de sus fases de desarrollo, destinados a la acuacultura y ornato en los estados unidos mexicanos. *Diario Oficial de la Federación*. 20 de Julio, 1994.

Ortiz, G. C. 1997. Esfuerzos que se desarrollan para la difusión de la acuariofilia. En: *Primer Congreso Nacional Acuariofilia*. Dirección General de Acuacultura. SEPESCA. México.

Ramírez-Martínez, C 1999. "La perspectiva de la acuacultura social en México" En: Memorias del IV encuentro de empresas sociales de pesca y acuacultura. *Secretaría de Desarrollo Social*. México.

Ramírez-Martínez, C. y Mendoza, R. 2005." La producción y comercialización de peces de ornato de agua, como vector de introducción de especies acuáticas invasivas en México" *En: Memorías del 37° Symposium del Desert Fishes Council*. Cuatrociénegas, Coah.

Rixon, C., I. Duggan, N. Bergeron, A. Ricciardi, and H. Macisaac. 2005. Invasion risks posed by the aquarium trade and live fish markets on the Laurentian Great Lakes. *Biodiversity and Conservation* 14: 1365-1381.

SAGARPA (Secretaría de Agricultura, Ganadería, Pesca y Alimentación). 2005. Delegación Estatal en el estado de Morelos http://www.sagarpa.gob.mx/dlg/morelos/pesca /informacion.htm

SEPESCA (Secretaria de Pesca). 1984. Reporte sobre el desarrollo de la acuacultura de subsistencia en Morelos. *Reporte Técnico*. Dirección General de Acuacultura. México, D.F. 45 p.

Taylor,J. N., W. R. Courtenay, Jr. and J. A. McCann. 1984. Knows impacts of exotic fishes in the continental United States. In: W. R. Courtenay, Jr. and J. R. Stauffer, Jr., eds. *Distribution, biology, and management of exotics fishes*. The Johns Hopkins University Press, Baltimore, MD. USA.

Tlusty, M., 2002. The benefits and risks of aquacultural production for the aquarium trade. *Aquaculture* 205: 203– 219

Torres-Orozco, B. R. y Kobelkowsky, A. 1991. *Los Peces de México*. AGT Editor, S.A. México, D.F. 235 p.

UNEP (United Nations Environment Programme). 2008. Monitoring of International Trade in Ornamental Fish. European Commission. Cambridge, UK 43 p.

Wabnitz, C., Taylor, M., Green, E., Razak, T. 2003. *From Ocean to Aquarium*. UNEP-WCMC, Cambridge, UK. 64 p

Welcomme, R. 1992. Pesca Fluvial. *FAO Documento Técnico de Pesca*, Roma.303 p.

Weigle, S. L. D. Smith, J. T. Carlton and J. Pederson. 2005. Assesing the risk of introducing exotic species via the live marine species trade. *Conservation Biology*. 19(1): 213-223

Williams, M. J. 1997. Aquaculture and sustainable food security in the developing world. Pages 15-51. In: John E. Bardach, editor. *Sustainable Aquaculture*. John Wiley and Sons, Inc., New York, NY.

Wijkström, U., A. Gumy and R. Grainger. 2002. *The State of World Fisheries and Aquaculture (Sophia) FAO*, 150pp

Whittington, R. J. and R. Chong. 2007.Global trade in ornamental fish from an Australian perspective: The case for revised import risk analysis and management strategies. *Preventive Veterinary Medicine*. (81)1-3: 92-116

INDEX

A

B

C

D

G

H

I

J

K

L

M

N

O

P

Q

R

S

T